T0211968

INTERNATIONAL CENTRE FOR MECHANICAL SCIENCES

COURSES AND LECTURES · No. 270

ARTERIES
AND
ARTERIAL BLOOD FLOW

EDITED BY
C.M. RODKIEWICZ
UNIVERSITY OF ALBERTA

SPRINGER-VERLAG WIEN GMBH

This volume contains 199 figures.

Originally published by Springer-Verlag Wien New York in 1983

ISBN 978-3-211-81635-6   ISBN 978-3-7091-4342-1 (eBook)
DOI 10.1007/978-3-7091-4342-1

PREFACE

With the ever increasing quality and volume of human knowledge we also arrive at a better understanding of our own health problems. This is particularly true in medical and paramedical research. Since diseases and defects of the human cardiovascular system remain one of the most important causes of troubles and death, researchers are expanding considerable energy to understand this complex system.

Late Professor W. Olszak recognized the importance of this branch of learning and was instrumental in the organization of the International Seminar on "Engineering and Medical Aspects of the Arterial Blood Flow" in Udine, of which this book is part of.

The book contents provide an exposition of the standard concepts and their application as well as provide some recent research and its results. The material is arranged in five chapters.

Chapter I concerns the blood rheology and its implication in the flow of blood. This is studied as dominated by plasma viscosity, hematocrit and red cell properties, namely aggregability and deformability. Quantitative models of highly concentrated suspensions, which exhibit shear-thinning, thixotropy and viscoelasticity are discussed. Annular (two phase) flow models are developed for analyzing blood flow in narrow vessels. Some examples in clinical application are given.

Chapter II deals with the arterial walls. The geometry and mechanical properties of blood vessels of the cardiovascular system are discussed in detail. Mathematical descriptions are presented and compared with experimental work. The time dependence is treated both in terms of quasilinear elasticity, continuous relaxation spectra, and general nonlinear viscoelasticity theory.

Chapter III briefly introduces the dynamics of fluid filled tubes. Initial value problems are considered for fluid filled elastic and viscoelastic tubes. Material properties are chosen which are appropriate for biological applications. Numerical techniques for the inversion of integral transforms, which arise in the analysis, are discussed and numerical results are presented.

Chapter IV concerns the small arteries and the interaction with the cardiovascular system. This chapter describes and defines the properties of the microcirculation: the *physical properties of the wall, the contractibility, the flow and pressure in the small vessels and the transcappillary movement of the fluid. Special parts of the circulation (particularly of coronary circulation) and the interaction between small arteries and the other components of the circulation are described.*

Chapter V deals with the flow in large arteries. Flow characteristics and governing parameters for the Newtonian fluids are discussed. Basic differential equations and shear stress expressions are set. This is followed with theoretical and experimental considerations pertinent to the flow in straight tubes, curved passages, simple junctions, and the aortic arch. Correlation with the atherosclerotic formations is indicated.

Czeslaw M. Rodkiewicz

CONTENTS

IN MEMORIAM

WACLAW OLSZAK

Prof. Olszak received Ph. D. degrees in Engineering and Mathematics at the Universities of Paris and Vienna.

In 1952, he founded the Department of Mechanics and Continuous Media, which has since become the most important and dynamic research center in Poland. Later he became head of that Department. His research problems focused on the theory of elasticity and plasticity, rheology and the theory of structures. He was also the first in Poland to tackle the problems of the theory of prestressed structures, and the results of his investigations were published in two monographs.

Prof. Olszak also authored more than 300 scientific and technical papers, as well as 10 monographs, and was the editor-in-chief of several journals of the Polish Academy of Sciences.

From 1936 till his death in 1980, Prof. Olszak participated in 5 international congresses at which he presented his original research papers. In addition, he organized several lecture tours in Canada and the United States, as well as in several countries in Europe.

His international reputation and authority in the field of plasticity has earned him numerous honorary degrees all over the world. Prof. Olszak had received the title of "Doctor Honoris Causa" from the Universities of Toulouse, Liège, Glasgow and New Brunswick, from the Technical Universities of Vienna, Dresden, Warsaw and Cracow and from the Mining College in Cracow.

In addition, he held membership at the Academies of Sciences in Toulouse, Belgrade, Stockholm, Helsinki, Turin, Halle, Vienna, Sofia, Boulogne, Paris, Madrid, Ljubljana and Buenos Aires, honorary memberships in Budapest and Luxemburg, and a Fellowship in New York.

During his career, Prof. Olszak founded many organizations all over the world. One of his more noteworthy accomplishments has been participating in the foundation of the International Center for Mechanical Sciences (CISM) in Udine, Italy, in 1969. The CISM is, so far, the only institution which, on an international level, covers all domains of Theoretical and Applied Mechanics. It has rapidly acquired a recognized international reputation. To a great extent, this was due to Prof Olszak, who as CISM's rector, engaged his scientific reputation and international experience to the benefit of this dynamic institution.

Czeslaw M. Rodkiewicz

BLOOD RHEOLOGY AND ITS IMPLICATION IN FLOW OF BLOOD

Daniel QUEMADA
Laboratoire de Biorhéologie
et d'Hydrodynamique Physicochimique
UNIVERSITE PARIS VII - 2, Place Jussieu
75251 Paris Cedex 05 (FRANCE)

SUMMARY : Rheological behaviour of blood is studied as dominated by
plasma viscosity, hematocrit and Red Cell properties, namely aggregability and deformability. Quantitative models for highly concentrated suspensions, which exhibit shear thinning, thixotropy and viscoelasticity,
are discussed. Annular (two-phase) flow models are developped for analysing blood flow in narrow vessels. Some examples in clinical application
are given.

1 - GENERAL FEATURES OF BLOOD CIRCULATION IN NARROW VESSELS.
   BASIC RHEOLOGICAL CONCEPTS.

1.1 Introduction

     Blood flow has been early considered as having a major importance in
physio-pathology : for a long time, blood remain as the vital fluid,
heart failure as death.

     Transport of oxygen and nutrients that tissue cells need for their
metabolic activity on the one hand and removal of carbon dioxide and
metabolic products, on the other hand, are the main functions of blood
circulation. Moreover under varying external conditions, peripheral (skin)
microcirculation maintains a stable temperature within the body.

     The rate of this exchange mainly depends on (i) red blood cell
(RBC) concentration (i.e. hematocrit H) in circulating blood and (ii)
flow rate Q. It is therefore determined by the cardiac performance. For
a long time, the latter has been thought of as only related to the ability of heart to deliver the energy for blood pumping. It now appears more
and more that flow resistance dominates the cardiac performance, not
only through characteristic properties of blood vessels (non linear (viscoelastic)mechanical properties ; geometrical factors - branching, tapering,
singularities - and vaso-control effects) but also through fluid properties. Indeed, rheological characteristics of blood (non-newtonian viscosity, thixotropy and visco-elasticity) and physico-chemical properties of
plasma, play a significant role in flow regulation in health and diseases.

     Several authors , as S. CHIEN (1972) stressed that "hemorheological
factors were unjustly underestimated up to now, in comparison with cardiac or vascular factors".
Indeed, it is not unreasonable to think that number of cardiovascular diseases — currently attributed to some myocardiac or vascular pathologies
which would appear in response to direct action of some external factor—
could be in fact a secondary effect resulting from a primary hemorheolo-
gical disease due to this external factor acting directly on blood.

One goal of hemorheology is the analysis of direct effects that many substances have on mechanical properties of blood and then, the study of the consequences that these modifications have on blood flows. Clinicians and Physiologists, and therefore, Hydrodynamicists have to begin their analysis with the study of these flow properties of blood. Before classifying them, let us describe briefly the main characteristics of blood.

1.2 Constituents of blood.

A. Blood composition.

Blood is a very highly concentrated suspension (about 40 - 45 % in volume fraction) of a variety of cells, Red Cells (RBC), White Cells (WBC), Platelets (°) suspended in a continous phase called plasma.

Plasma is an aqueous solution of electrolytes and organic substances mainly proteins. Relative proportions of cell elements and concentrations of plasma components in blood are shown on Table 1.1. Plasma behaves as a newtonian fluid whose viscosity η_p depends primarily on temperature as water does. Strong non newtonian properties of plasma reported in some observations had been considered as resulting from air-plasma interfacial

TABLE 1.1 - BLOOD CONSTITUENTS (5.10^6 Particles/mm^3)

CELL ELEMENTS	(Relative proportions)
White cells (all kinds)	1
Platelets	30
Red cells	600

PLASMA	(Weight fraction)
Water	0,91
Inorganic solutes	0,01
Proteins	0,07
Other organic substances	0,01

(°) alternatively called, erythrocytes, leucocytes and thrombocytes, respectively.

effects and therefore must be considered as non-existing. However, it can be expected, that due to the presence of long macromolecules, plasma should exhibit non-newtonian properties at low shear rate and that using a guard ring is required to avoid the formation of an interfacial film of proteins.

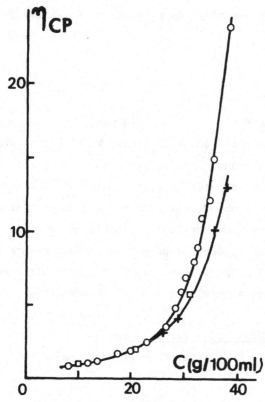

Fig.1.1. Viscosity of RBC content (from Data of CHIEN et al., 1970).

B. General properties of blood.

As RBCs compose about 97% of the total cell volume, whole blood properties are (rheologically) dominated by RBC properties (excepted in some pathologic cases as leukaemia in which the large number of white cells increases both viscosity and shear-rate dependence of blood (DINTENFASS, 1965)).

Hematocrit is the volume fraction of red cells plus trapped plasma obtained by centrifuge of a volume of whole blood prevented from clotting by addition of anticoagulants. Blood rheological properties have

been found free from any effects that   would be induced by these antico-
agulants(generally heparin or EDTA (°)). The hematocrit value H is normally a little greater than the true volume fraction φ , since a small volume of plasma is trapped between the cells and it is advisable to use a
corrected hematocrit φ = 0.96 H in normal conditions (CHIEN and USAMI,
1971). Nevertheless, this correction strongly depends on RBC deformability and becomes very important in the case of rigid particles for which
packing volume fraction is about 0.60. Such a value for the correction
factor was found in the case of hardened cell suspensions (CHIEN and
USAMI, 1971).

C. The red cell.

   Human RBC can be considered as a partially inflated balloon
(BROCHARD, 1977), thus having a very high level in deformability, which
decreases with the age of the cell. Internal fluid is a hemoglobin solution (35 g/100 ml), the viscosity of which is about 6 to 8 cP (See
Fig. 1.1) . RBC membrane consists of a biomolecular leaflet (a lipid
bilayer, with inclusion of anchored proteins) supported by a more rigid
skeletal structure (the actin-spectrin network). RBC membrane carries a
negative electrical charge, equivalent to about 6.000 electronic charges
per cell.
   The RBC biconcave discoid shape (in the case of mammals) results
from the ejection by the cell of its nucleus on entering the circulation. This leads to a volume reduction of the cell, its area remaining
constant. RBC dimensions (Fig.1.2) are greater than the critical ones for
which Brownian motion becomes important. Therefore, under steady conditions, blood settling occurs. After one hour, a plasma column above the
settling RBC is formed, the height of which is about 1 cm for normal
blood, but which can be ten (or more) times higher in diseases (in hos-

----

(°) EDTA = disodium-diamino-ethane-tetra-acetate.

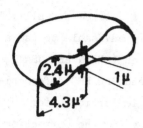

Fig.1.2 Human RBC dimensions

pitals, this height is currently used as a clinical index, the ESR (°) Although these sedimentation effects can be neglected when blood flows very rapidly, they are more and more important as flow rate is decreasing. Changes in osmotic conditions (normal ones correspond to those of an aqueous solution containing about 0.9% by weight of NaCl) induce drastic changes in RBC shape. Indeed, RBC becomes spherical (and then is called *spherocyte*) after swelling in solutions with lower NaCl concentration than normal ones. Further swelling leads to hemolysis of the cell which eliminates hemoglobin towards the plasma, the result being a *ghost*, i.e. a RBC reduced to its membrane only. Under higher NaCl concentrations, on the contrary, the RBC volume decreases, what leads to the formation of *echinocytes*, the membrane structure of which **likely involves the actin-spectrum network.**

D. Plasma proteins and coagulation

Plasma proteins (See Table 1.1) are mainly albumins and globulins which control water exchanges between blood and tissues by their action on osmotic balance. The remainder **are** lipoproteins and fibrinogen. The latter plays a very important role in the clotting mechanism of blood. However, under normal conditions, and in association with β-globulins, fibrinogen is closely involved in the *reversible aggregation* of RBC which form *rouleaux*, Fig.1.3. We will see later that it is one of the fundamental determinants of blood rheology.

Whole blood solidifies in the presence of air oxygen, or coming

----

(°) Erythrocyte Sedimentation Rate

into contact with extracorporeal surfaces. The formation of a clot would
be observed after an early increase in viscosity (DINTENFASS, 1971). In
vitro, this clot mainly consists of a complex structure of RBC and fi-
brin which derives from the plasma fibrinogen by a chain reaction. Platelets play a predominant role in clotting under in vivo conditions.
After complete formation, a contraction of the clot (COPLEY, 1960) begins,
which resembles *syneresis*, a similar retraction observed during gelation
process in colloids. After the clotting is achieved, the extra-fluid
is called *serum*, which is roughly plasma without fibrinogen.

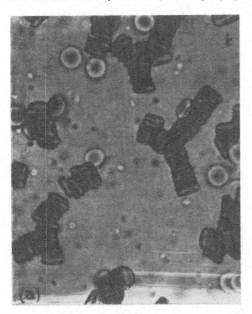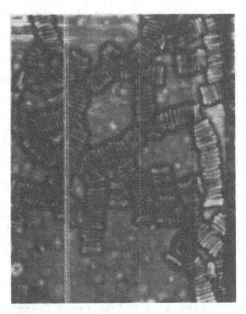

Fig.1.3 Photomicrographs of rouleaux of
human RBC (a) which tend to form a more
or less complicated network (b).

1.3. Main factors in blood circulation and continuum model

These blood properties play different roles as one considers different parts of the body.

Three main classes of blood circulation are usually sorted out, according to the ratio of the vessel diameter, 2R, to the particle "size",
2a, $\xi = R/a$. These groups are the following:

a) The capillary circulation (°) ($\xi \lesssim 1$) in a very large number (1.2×10^9) of very narrow (3 µm $\lesssim$ 2R $\lesssim$ 8 µm) and relatively long ($\ell \sim$ 1mm) vessels. Inside the vessels, RBCs undergo a strong deformation. Hence, the capillary circulation is dominated by *RBC rheological proper-ties* i.e. the membrane visco-elasticity and the internal fluid viscosity.

b) The systemic circulation ($\xi \gtrsim 50$) in few thousands of arteries and veins (0.6 mm $\lesssim$ 2R $\lesssim$ 20 mm; 1 cm $\lesssim \ell \lesssim$ 40 cm), inside which blood flow pulsatility promotes important deformations of vessel walls. Thus this circulation involves the *rheological properties of vessel walls -* i.e. their non linear viscoelasticity - related to their microscopic structure, with elastin, collagen and smooth muscle as the three major wall materials (BERGEL, 1964). Although it was currently accepted that in large vessels blood behaves as a newtonian fluid, it seems that, es-pecially in veins, abnormal (i.e. non-newtonian) properties of blood must be taken into account (FLAUD and QUEMADA, 1980).

c) The microcirculation (°) (3 $\lesssim \xi \lesssim$ 50) in arterioles and venules (about 10^8 in number; 10 µm $\lesssim$ 2R $\lesssim$ 500 µm ; 2 mm $\lesssim \ell \lesssim$ 10 mm) which is governed by these *anomalous (non-newtonian) blood properties*. Such pro-perties are dominated by blood microscopic structure and associated with the formation of a *particle depleted wall layer* (the so-called plasma layer) and *a particle rich axial core*. The existence of such a two phase (annular) flow leads to "anomalies" as Farhaeus Effect and Farhaeus-Lindquist Effect.

<u>Limits of Hydrodynamic Description</u>.

It must be noticed that this classification corresponds closely to the limits in validity of the description of the fluid as a quasi-homo-geneous media, hence in validity of the hydrodynamic approximation. In

----

(°) Number of authors used the term microcirculation for circulation both (i) in capillaries and (ii) arteriols and venules. Hereafter we prefer to limit the use of this to the latter.

§ 1.3a, the problem is confined to the mechanics of deformable bodies im-
mersed in a newtonian fluid (the plasma), in the presence of rigid walls.
In § 1.3b, as a <<<R, blood can be considered as a homogeneous fluid, having an effective (bulk) viscosity, more or less newtonian, (which however depends on concentration). In § 1.3c, the problem is much more complex
since hydrodynamic approximation does not hold at this scale. Nevertheless,
on the artery (or vein) side, the plasma layer thickness δ remains small
in comparison with vessel radius and the validity of the hydrodynamic approximation is recovered when assuming that the actual flow can be described as an *annular two-phase flow* of two different quasi homogeneous
fluids. As the ratio ξ = R/a decreases, the previous axial core, which
undergoes shear stresses (°) is progressively replaced by a more or less
axial single file of cells, in which cells follow each other keeping their
discoidal surfaces approximately perpendicular to the axis of the vessel.
Such a structure, named *axial train* by R.L. WHITMORE (1967), is still a
two-phase structure and it can be thought that, at least approximatively,
a continuous description by such an annular two-phase flow would be possible down to ξ →1 .

   Anyway, using such two-phase flow models requires the knowledge of
bulk rheological properties of blood, which will be surveyed briefly after
recalling some basic rheological principles and experimental methods. Then,
general (more or less empirical) models will be studied with hope that
they may provide a basis for experimental and clinical data interpretation
of blood flow in narrow vessels (in vitro and, as far as possible, in vivo).

1.4. Basic rheological concepts.

   A) Viscometric flows

   As rheological behaviour of complex materials is very complicated,
the simplest flow conditions were used for their rheological characteri-

----

(°) This annular flow can be called a sheared-core.

zation. These conditions are achieved in *viscometric* flows, as *simple shear* (which occurs in shear and pressure flows restrained by fixed boundaries i.e. Couette and Poiseuille flows) or as *pure shear* and *axisymmetric distorsion* (which occurs in elongational or compressional flows, associated with extension of sheets or filaments, respectively).

B) Steady shear flows : shear viscosity.

B.1 Newtonian viscosity.

Uniform plane shear is the flow obtained when two flat parallel plates with fluid in between are moved uniformily past one another, the distance h between them being kept constant and the fluid being assumed to adhere to both plates (Fig.1.4). A force per unit area (shear stress

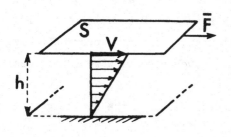

Fig.1.4 Shear flow between two parallel plates

$\sigma = F/S$ in dynes/cm^2) is required to move one plate (the other being stationnary) with the velocity V. In ordinary fluids and under steady conditions, the fluid velocity varies linearly from $v = o$ (on the stationary plate) to $v = V$ (on the moving plate). One gets a plane Couette Flow, which approximates the Couette Flow between coaxial cylinders (Radius R_1 and $R_2 = R_1 + \epsilon$) when rotating the outer one, if the gap width ϵ is very small in comparison with $R_1 \simeq R_2$. The (constant) velocity gradient (or shear rate, in sec^{-1}) is $\dot{\gamma} = V/h$.

a) For number of (ordinary) fluids, as water, oils, simple organic fluids ..., the shear stress σ is found varying linearly with the shear rate (Newton law):

$$\sigma = \eta\dot{\gamma} \qquad\qquad\qquad (1.1)$$

the(constant) coefficient being the viscosity (in Poises):

$$\eta = \frac{\sigma}{\dot{\gamma}} \tag{1.1a}$$

Fluids to which (1.1) applies are called Newtonian fluids, the viscosity of which is constant, i.e. does not depend on $\dot{\gamma}$ (or σ).

Furthermore, solutions and disperse systems (as suspensions, emulsions, ...) at least inside a limited range of shear rate, show a newtonian viscosity which depends on particle concentration (See Fig.1.5 from GOLDSMITH, 1973).

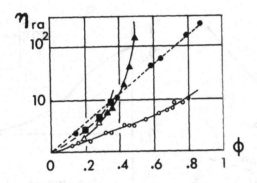

Fig.1.5 Relative apparent viscosity η_{ra} of particle suspensions, as a function of volume fraction ϕ. (Human blood O; de-oxygenated sickel cell ■ ; rigid spheres ▲ ; rigid discs Δ ; droplets ● ; (From GOLDSMITH,1973).

B.2 Non newtonian viscosity.

However, any number of usual "fluids", as polymer solutions and melts, inks, paints, slurries, colloïds, bio-fluids ... deviate from Newton law. Therefore, their "viscosity", as the ratio of stress to rate of shear (1.1a) is not a constant, but depends on σ or $\dot{\gamma}$.

$$\sigma = \sigma(\dot{\gamma}) \qquad \text{or} \qquad \dot{\gamma} = \dot{\gamma}(\sigma) \tag{1.2}$$

$$\eta = \frac{\sigma}{\dot{\gamma}} = \eta(\dot{\gamma}) \qquad \text{or} \qquad \eta(\sigma) \tag{1.3}$$

Such fluids are called non-newtonian fluids, in which additional forces arise from anisotropy of internal pressure. For the plane shear, these forces called normal (or Weissemberg) forces, acting in the direction
normal to the streamlines, tend to separate the plates. In most cases the
viscosity (1.3) decreases continuously when $\dot{\gamma}$ (or σ) increases: the fluid
is then called a *shear-thinning* (or *pseudoplastic*) fluid. Fig. 1.6 illustrates such a variation.On the contrary, it is called a shear-thickening
fluid,sometimes related to dilatant effects observed in nearly close-
packed disperse systems, as water-sand mixtures.

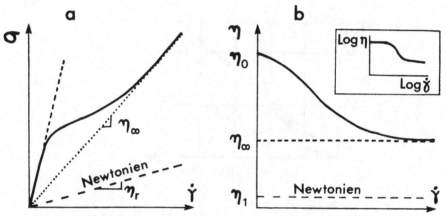

Fig.1.6 Non-newtonian (Shear-Thinning) Steady Behaviour a) $\sigma=\sigma(\dot{\gamma})$
variations; b) Structural viscosity $\eta=\eta(\dot{\gamma})$, having the two limiting
values $\eta_0=\eta(o)$ and $\eta_\infty = \eta(\infty)$:(On set, curve η versus $\dot{\gamma}$, in logarithmic coordinates, exhibits an apparent plateau for $\dot{\gamma} \to o$).

As most of these fluids are disperse systems, their pseudoplastic behaviour has been attributed to structure modifications of the system under
shear stresses. For instance, at very low shear rate, particles form *large clusters* (or flocs, or aggregates), sometimes leading to a *tri-dimensional network*, like in a gel. Such structures are progressively broken
down by shear stresses, as shear rate increases, which leads to the corresponding reduction of viscosity. At a given steady shear rate, a dynamic equilibrium between formation and destruction of aggregates is to be
observed. If particles are not rigid, their deformation grows by increa-

sing the shear rate, that leads to increase the particle orientation in
the flow direction (the same effect is obtained with non spherical particles). Hence, an additional cause of lowering of viscosity as $\dot{\gamma}$ increases
is found out. This discussion calls for the very important concept of
structural viscosity (first introduced in 1926 by OSTWALD and AUERBACH).

c) Steady elongational (or extensional) flow.

   This type of flow (also called pure straining motion) corresponds to
the velocity field generated when an incompressible fluid is drawn out
into threads or sheets. For instance, for a thread parallel to the x-coor-
dinate the velocity components are, with $\kappa = \partial v_1 / \partial x_1$, $v_1 = \kappa x$,
$v_2 = -\frac{1}{2}\kappa x_2$, $v_3 = -\frac{1}{2}\kappa x_3$, and the material behaviour under this uniaxial
flow is described by the elongational (or Trouton) viscosity η_E such as:

$$\sigma_{11} - \sigma_{22} = \sigma_{11} - \sigma_{33} = \eta_E \kappa \qquad\qquad (1.4)$$

   For Newtonian fluids, $\eta_E = 3\eta$, η being the (simple) shear viscosity. However, $\eta_E = \eta_E(\kappa)$, for non-newtonian fluids, and no relation between $\eta_E(\kappa)$ and $\eta(\dot{\gamma})$ can be predicted. Large enhancements of η_E in suspensions of elongated particles when compared to η_E of the suspending
fluid, have been experimentally observed (WEINBERGER and GODDARD 1974).
These enhancements are much larger than the corresponding increases in
shear viscosity η .

   Although η_E is more difficult to measure than η and only few data
are thus available, such extensional effects could be of some importance
on blood circulation since extentional flows exist, at least somewhat
partly, at branching of vessels.

d) Unsteady shear flows: thixotropy and viscoelasticity

   The existence of a shear dependent internal structure leads to *time-
dependent effects*, since structure modifications resulting from a sudden
change of shear rate (or stress) are not instantaneous.

d 1) The response of a shear thinning material to the application at t = 0 of a sudden step of shear rate (from 0 to $\dot{\gamma}_1$ for instance) is shown on Fig. 1.7 . At t = 0, the system behaves without changing the structure that it had at zero shear rate, i.e. it behaves like a fluid

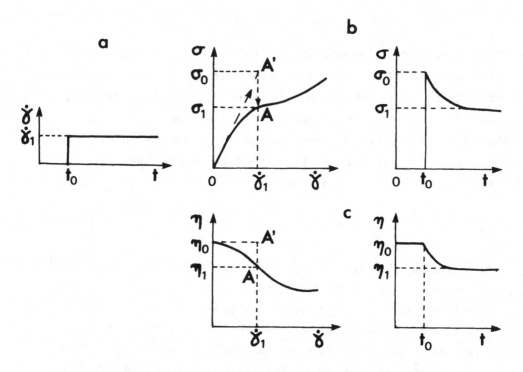

Fig.1.7 Variations of $\sigma(t)$ and $\eta(t)$ for a thixotropic fluid submitted to a step of shear rate $\dot{\gamma}$, in relation with shear rate dependence of steady shear stress and viscosity, $\sigma(\dot{\gamma})$ and $\eta(\dot{\gamma})$.

having a viscosity $\eta_0 = \eta(\dot{\gamma} = 0)$ (or like a gel if the structure at $\dot{\gamma} = 0$ is a tri-dimensional network of aggregates, with $\eta_0 \sim \infty$). Thus, at t = 0, the shear stress instantaneously reaches the value $\sigma_0 = \eta_0 \dot{\gamma}_1$ (from 0 to A', Fig.1.7b). However, the zero shear structure cannot be maintained at $\dot{\gamma} = \dot{\gamma}_1$, and it relaxes (from A' to A) towards the equilibrium structure (i.e. the dynamic equilibrium at $\dot{\gamma} = \dot{\gamma}_1$), the viscosity of which being $\eta_1 = \eta (\dot{\gamma} = \dot{\gamma}_1)$ and the corresponding shear stress,

$\sigma_1 = \eta_1 \dot{\gamma}_1$. This structure adaptation involves relaxation times, which characterize the relaxation of the structure. Such a time behaviour is called
thixotropy (Fig.1.8c) for shear-thinning materials, and rheopexy for shear
thickening materials.

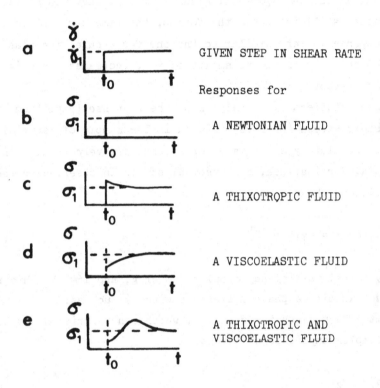

Fig.1.8 Different types of non-newtonian behaviour under a step
of shear rate.

If the shear stress – shear rate relaxation $\sigma = \sigma(\dot{\gamma})$ is not linear,
the type of time curve $\sigma(t)$ depends on the step height $\dot{\gamma}_1$. At low $\dot{\gamma}_1$
values, the response $\sigma(t)$ is very similar to the one of a newtonian
fluid (Fig.1.8b). As $\dot{\gamma}_1$ increases, the difference $\sigma_0 - \sigma_1$, between the
peak of the curve σ_0 and the equilibrium value σ_1, grows.

d2) Thixotropy is not the only time behaviour currently observed. A
very large number of materials exhibit a *viscoelastic behaviour*, associa-

ted with a purely elastic (instantaneous) response, followed by a delayed
one (retarded elasticity) (Fig.1.8d). Nevertheless, the latter is very
often superposed to thixotropy and it appears very difficult to separate
these two types of effects (which moreover would be overshadowed by the
instrument dynamics). A typical time-curve $\sigma(t)$ is shown on Fig.1.8e.
Non-linearities still influence the form of the time curve: at very low
$\dot{\gamma}_1$, the response is very similar to the one for a pure viscoelastic (as
in Fig.1.8d) and increasing the amplitude $\dot{\gamma}_1$, increases the height of
the peak ("overshoot").

Dynamical studies, i.e. analysis of the response (amplitude and
phase) in shear stress (or rate) induced in the sample of material under
applying an oscillatory shear rate (or stress) are very usual in viscoelastic media. For instance, a sinusoidal stress in observed in response
to an oscillatory shear rate:

$$\dot{\gamma} = \dot{\gamma}_1 \cos\omega t = \dot{\gamma}_1 Re(e^{i\omega t}) \qquad\qquad (1.5)$$

where $Re(g)$ = real part of any complex number g, and $i^2 = -1$. This response in stress exhibits a phase difference with $\dot{\gamma}$ such as:
$\sigma = \sigma_0 \cos(\omega t + \varphi) = \sigma_1 \cos\omega t + \sigma_2 \sin\omega t$, which can be written $\sigma = Re(\tilde{\sigma}e^{i\omega t})$
using the complex amplitude of stress:

$$\tilde{\sigma} = \sigma_1 - i\sigma_2 \qquad\qquad (1.6)$$

This leads to introduce a *dynamical viscosity* as a complex number

$$\tilde{\eta} = \eta_1 - i\eta_2 \qquad\qquad (1.7)$$

where $\eta_1 = \dfrac{\sigma_1}{\dot{\gamma}_1}$ and $\eta_2 = \dfrac{\sigma_2}{\dot{\gamma}_1}$ are the viscous and elastic components, respectively.

2 - VARIABLES GOVERNING SUSPENSION RHEOLOGY - BRIEF SURVEY OF EXPERIMEN-
    TAL METHODS.

2.1 Variables governing mechanical behaviour of disperse systems : Di-
    mensional approach to suspension rheology.

Attempt to predict flow behaviour of concentrated suspensions require to introduce "structural parameters" into rheological equations in
order to interrelate the rheological properties of the material and its
molecular or particulate structure. Progress in choosing such parameters
can be gained using a dimensional approach similar to that given by
KRIEGER (1963) for colloid suspensions.

The simplest system (from the structural point of view) is a suspension of uniform rigid spheres in a newtonian fluid. For non-interacting spheres, the viscosity η of this suspension will depend on the
eight following variables :

  o <u>particle variables</u> :   (dimension)

     sphere radius a (L)

     density ρ_P (ML^{-3})

     number per unit volume n (L^{-3})

  o <u>suspending fluid variables</u> :

     viscosity η_F $(ML^{-1}T^{-1})$

     density ρ_F (ML^{-3})

     thermal energy KT $(ML^{-2}T^{-2})$

    (K being the Boltzman Constant)

  o <u>flow variables</u> :

     shear rate $\dot{\gamma}$ (T^{-1})

     "experimental" time t (T)

     (or period) $(\tau = f^{-1} = 2\pi\omega^{-1})$

With η, these nine variables can be combined to form six dimension-less groups :

- the relative viscosity $\qquad\qquad \eta_r = \eta/\eta_F$ $\qquad\qquad$ (2.1)
- the volume fraction $\qquad\qquad \phi = (4\pi/3)\, na^3$ $\qquad\qquad$ (2.2)
- the relative density $\qquad\qquad \rho_r = \rho_p/\rho_F$ $\qquad\qquad$ (2.3)
- the internal Reynolds number $\quad Re_i = \rho_F a^2 \dot\gamma/\eta_F$ $\qquad$ (2.4)
- a reduced shear rate $\qquad\qquad \dot\gamma_r = \tau\dot\gamma$ $\qquad\qquad$ (2.5)
- a reduced time $\qquad\qquad\qquad t_r = t/\tau$ $\qquad\qquad\qquad$ (2.6)

In (2.5) and (2.6), τ is a characteristic (internal) time of the system. Taking τ as the Brownian diffusion time

$$\tau \sim a^2 D_{tr}^{-1} \sim D_{rot}^{-1} \qquad\qquad (2.7)$$

where $D_{tr} = KT/6\pi\eta_F a$ and $D_{rot} = KT/8\pi\eta_F a^3$ are the translational and rotational Brownian coefficients for diffusion, respectively, leads to

$$\tau = b\eta_F a^3/ KT \qquad\qquad (2.8)$$

where b = numerical constant. Then (2.5) appears as closely related to translational and rotational (internal) Peclet numbers, $Pe_t = a^2\dot\gamma/D_{tr}$ and $Pe_r = \dot\gamma/D_{rot}$, respectively and (2.6), the inverse of the Deborah num-ber, De.

Using these reduced groups leads to the general form of the relati-ve viscosity equation :

$$\eta_r = \eta_r (\phi,\ \dot\gamma_r,\ t_r\ ;\ \rho_r,\ Re_i) \qquad\qquad (2.9)$$

In the case of neutrally buoyant particles, ($\rho_r=1$) and assuming la-minar Stokes flow ($Re_i \ll 1$), (2.9) reduces to :

$$\eta_r = \eta_r \ (\phi, \ \dot{\gamma}_r \ , \ t_r) \tag{2.10}$$

Steady laminar flow at fixed $\dot{\gamma}_r$ gives $\eta_r(\phi)$, that will be studied in the next § 3. Steady laminar flow at fixed ϕ leads to $\eta_r(\dot{\gamma}_r)$, the variation of which has been given on Fig. 1.6b for a shear-thinning material, for instance. Nevertheless, in the case of very large particles, as glass beads, τ is very high $[\tau_{(sec)} \simeq 6 \ b \ (a_{(\mu m)})^3]$. Thus, even for the smallest value available in viscometry, $\tau >>> 1$: a constant viscosity (close to the high shear limit) is observed, i.e. the fluid appears as newtonian. Likewise, $\tau <<< 1$ for very small particles, as molecules of a low molecular weight solute, giving again $\eta_r = c^{te}$ (but equal to the low shear limit), i.e. a newtonian behaviour.

Unsteady experiments as stress recovery after sudden stop of the viscometer (See § 1.4D) will involve the same characteristic time τ (as a relaxation time for thixotropic buildup of the structure at $\dot{\gamma} = 0$) since $\eta_r = \eta_r(t_r)$ in this case.

"Turbulence effects"can be occur if Re_i becomes larger than a critical value Re_i^{cr}. In fact, critical   tube Reynolds number, $\rho VR/\eta$ is found at least two orders in magnitude lower than the usual critical value for ordinary fluids.

In the case of more complex systems, each additional variable requires a new dimensionless variable in the viscosity equation, as shown in the following examples. For a suspension of ellipsoidal particles, the axial ratio is $a_\parallel/a_\perp$ of semi axes of rigid spheroid. Purely elastic particles (with a shear modulus G) suspended in a fluid of viscosity η_F have a relaxation time $(^\circ)$ $\tau_m = \eta_F/G$ that will be used for large defor-

----

$(^\circ)$ that is the Maxwell relaxation time of a viscoelastic liquid (which shows how viscosity and elasticity mix to give visco-elastic properties) (See § 6).

mable particles (since τ in (2.8) is too large) to define $\dot{\gamma}_r = \eta_F \dot{\gamma}/G$, as
the ratio of shear force to the elastic one. An emulsion involves two
additional parameters, the interfacial tension Γ and the internal viscosity of droplets η_i that leads to the relaxation time $\tau_D = \eta_F a/\Gamma$
and to the new dimensionless variables $\tau_D \dot{\gamma}$ and η_i / η_F. Furthermore, accounting for particle interactions involves ratio W/KT of some characte-
ristic energy W to the thermal one : for instance, short range Van der
Waals interaction (energy $\sim A/6 \frac{r}{a}$, where A is the Hamaker constant)
leads to take as new variable the ratio A/KT; long range forces as electrostatic repulsion are governed by the surface potential (or surface
charge density) and by screening effects from the ionic double layer,
the thickness κ_D^{-1} of which (the so-called Debye length) enters in the
dimensionless group $\kappa_D a$.Comparing experiments performed at different
temperatures or ionic strengths allow estimation of effects that these
dimensionless parameters would have on the rheological behaviour.

2.2. Experimental methods.

   Rheological studies of non-newtonian fluids are at first the measurement of the shear stress-shear rate relationship under steady conditions.
Structural changes involve time-dependent measurements in which viscoelastic and thixotropic effects are mixed, leading to difficulties in data interpretation. Nevertheless, as some structure knowledge is expected
from rheological measurements, complementary experiments are often carried out to verify structural models introduced in rheological studies.
There follows a brief description of them.
(a) Microscopic observations under shear controlled conditions give information about (i) particle orientation and (eventually) deformation,
(ii) particle aggregation and their variations with shear rate. Quantitative approaches have been performed using photometric aggregometry
(KLOSE et al, 1972), or laser back-scattering (MILLS et al, 1980).

(b) Sedimentation under controlled shear rates (COPLEY et al.1975; VIN-
CENT et OLIVER, 1977) exhibits variations of ESR closely related to the
aggregability of particles. (c) Filtration experiments (on polycarbonate
sieves with different mean pore diameters) have been believed as measuring particle deformability (for ex. GREGERSEN et al., 1967) although
artefacts result from particle aggregates at the entrance of the pores.
(d) Doppler-Laser Velocimetry (for ex. BORN et al., 1978) allows us to
fit model velocity profiles on the experimental ones, giving some in-
sight into the plasma layer influence.

A brief survey of the operating principles and main characteristics
of usual viscometers will follow.

2.3 Viscometry

Steady measurements of blood rheological properties concern mainly
its shear viscosity. Two types of viscometers are currently used, capillary (tube) viscometers and rotational viscometers.

A. Tube viscometers

Fluid flows down a tube from a reservoir. The tube diameter 2R is
precisely known (nevertheless this diameter is always larger than about
300 μm, in order to avoid phase separation effects (See §7.1). One records the time interval during which a fixed volume of fluid flows
through the tube under a given pressure gradient $P = \Delta p/L$. As shear rate varies inside the tube, decreasing from its maximum value $\dot{\gamma}_m$ at the
tube wall to zero at the tube axis, the measurement involves the (unknown) flow curve between 0 and $\dot{\gamma}_m$. In the case of a newtonian fluid,
no difficulty arises since $\eta = c^{te}$ and the pressure-flow rate relation
(Poiseuille law) give the viscosity

$$\eta = \frac{\pi}{8} \frac{R^4}{Q} P \qquad\qquad (2.11)$$

Corrections for end effects have been taken into account using pairs of capillary tubes of equal radii and of different lengths (COKE-LET, 1972).

In the case of non-newtonian fluids, eq.(2.11) is used to define the *apparent viscosity of the suspension* η_a, which is the viscosity of a fictive newtonian fluid which would give the same flow rate under the same pressure gradient. In the case of homogeneous fluid (with local $\dot{\gamma}$ value as a single-valued function of local σ-value), the true relation-ship between $\dot{\gamma}$ and σ can be calculated from the values at the wall, σ_w and $\dot{\gamma}_w$:

$$\sigma_w = \frac{1}{2} RP \qquad \text{and} \qquad \dot{\gamma}_w = \dot{\gamma}(\sigma_w) \qquad\qquad (2.12)$$

Using the local shear rate $\dot{\gamma} = -dv/dr$, where $v(r)$ is the axial component of velocity, the flow rate expression can be integrated by parts :

$$Q = 2\pi \int_0^R v(r)\, r\, dr = -\pi \int_0^R r^2 \frac{dv}{dr}\, dr \qquad\qquad (2.13)$$

using the no-slip boundary condition at the wall $r = R$, $v(R) = 0$.

Changing the variable r into the local shear stress $\sigma = \sigma_w \frac{r}{R}$ trans-forms (2.13) into :

$$Q = \frac{\pi R^3}{\sigma_w^3} \int_0^{\sigma_w} \sigma^2 \dot{\gamma}\, d\sigma \qquad\qquad (2.13\ a)$$

After differention of (2.13a) with respect to σ_w and using (2.12), the Rabinowitsch-Mooney equation is obtained which gives the true visco-sity η (as the ratio $\sigma_w / \dot{\gamma}_w$) from the experimental data $Q = Q(P)$:

$$\frac{1}{\eta} = \frac{2}{\pi R^4} \left[\frac{dQ}{dP} + 3 \frac{Q}{P} \right] \qquad\qquad (2.14)$$

It is worth noting that (2.14) gives the value of the true viscosity $\eta = \eta(\dot{\gamma})$ for $\dot{\gamma} = \dot{\gamma}_w$, the true shear rate at the wall. Both η and $\dot{\gamma}_w$ differ respectively from η_a (by 2.11) and $\dot{\gamma}_{aw} = 4Q/\pi R^3$, the latter being the (fictive) wall shear rate of a fictive newtonian fluid (with viscosity η_a). Eq (2.11) is recovered if the fluid is assumed to be newtonian, i.e. if $dQ/dP = Q/P$.

Nevertheless numerical derivative dQ/dP is required (since Q and P are known as finite sets of values) and for the sake of simplicity results are often given using apparent viscosity η_a vs. apparent shear rate $\dot{\gamma}_{aw}$. Moreover enhanced entry and exit effects can occur with viscoelastic fluids, depending on the average time t_a of passage of the fluid through the tube. (One needs $t_a \gg \tau$, the relaxation time of the fluid, in order to reach steady flow).

B. Rotational viscometers

The two main types are the coaxial cylinder and the cone-plate viscometers.

B.1 - Coaxial cylinder viscometer (Fig. 2.1)

The fluid is held between a cylindrical (inner) bob (radius = R, height = h) and a coaxially mounted (outer) cup (radius = sR, s > 1). The cup is rotated at constant angular speed Ω and the viscous drag transmitted through the fluid is measured by the angular deflection of the bob (suspended from a torsion wire) or by the torque L required to maintain it at its original position.

When the gap between cup and bob is small enough (s - 1 $\ll$ 1), the mean shear stress σ_a and the mean shear rate $\dot{\gamma}_a$ are defined as :

$$\sigma_a = \frac{s^2 + 1}{s^2} \frac{L}{4\pi R^2 h} = \frac{s^2 + 1}{2s^2} \sigma_w \qquad\qquad (2.15)$$

$$\dot{\gamma}_a = \frac{s^2 + 1}{s^2 - 1} \, \Omega \qquad\qquad\qquad\qquad\qquad (2.16)$$

where $\quad \sigma_w = \dfrac{L}{2\pi R^2 h}$, is the shear stress on the bob surface.

True viscosity (for newtonian fluids) or apparent viscosity (for non newtonian fluids) is given by eq. (2.15) and (2.16) such as :

$$\eta_a = \frac{\sigma_a}{\dot{\gamma}_a} = \frac{s^2 - 1}{s^2 \, \Omega} \, \frac{L}{4\pi R^2 h} \qquad\qquad\qquad (2.17)$$

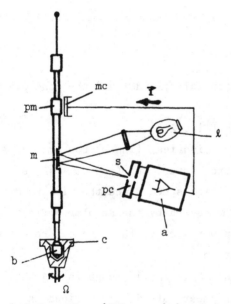

Fig.2.1 Coaxial cylinder
        viscometer.
The bob (b) undergoes shear
forces resulting from the rotation of the cup (c), driven
(angular speed Ω)by a motor
(not represented here).

Torque measurements are performed by recording the current $\vec{i}$
through the coils (mc) which
creates on the permanent magnet
(pm) an electromagnetic moment
opposite to the torque moment.
This current is generated by
the photo-electric cells (pc)
and amplifier (a) when the spot
s from the lamp (l) is reflected by the mirror (m) is devia-
ted from the middle position between the cells.

For non newtonian fluids (restricted to the case when local shear stress is an only function of local shear rate), the true viscosity can be obtained from the following infinite series (KRIEGER and ELROD, 1953) which gives the shear rate $\dot{\gamma}_W$ at the bob surface :

$$\dot{\gamma}_W = \frac{\sigma_W}{\eta} = \frac{\Omega}{\chi} \left[1 + \chi \frac{dy}{dx} + \frac{1}{3} \chi^2 \left(\frac{d^2y}{dx^2} + \left(\frac{dy}{dx}\right)^2 \right) + \dots \right] \qquad (2.18)$$

where : $\chi = \text{Ln } s$, $x = \text{Ln } \sigma_W$, and $y = \text{Ln } \Omega$

As in tube viscometers, end and edge effects can be eliminated using two bobs with different heights.

B2. Cone and Plate viscometers (Fig. 2.2)

The fluid fills the gap between a cone of very large apex angle 2θ (such as $\alpha = 90° - \theta \leq 4°$ for example) and a flat surface (plate) normal to the cone axis. Cone apex just rests on the plate. One unit (cone or plate) is rotated (angular speed Ω) and the viscous drag is measured as a torque L on the other. If phase separation does not occur the fluid inside the gap experiences a constant shear rate $\dot{\gamma} = \Omega/\alpha$. If R is the "cone radius" the shear stress at the plate surface is

$$\sigma_W = 3L/2\pi R^3 \qquad\qquad (2.19)$$

from which the apparent viscosity η_a (the true one for newtonian fluids) can be calculted directly :

$$\eta_a = \sigma_W/\dot{\gamma} = 3\alpha L/2\pi R^3 \Omega \qquad\qquad (2.20)$$

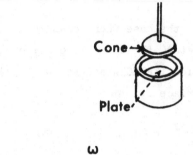

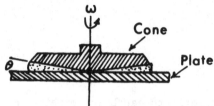

Eq. (2.20) approximates the true
viscosity within the limits of validity of constant shear rate $\dot{\gamma}$ inside the gap (since $\dot{\gamma}$ is very close
to $\dot{\gamma}_w$, the shear rate at the plate
surface).

Fig. 2.2 Cone/Plate viscometer.

C. Some experimental problems in viscometry.

Different kinds of problems arise in viscometry of disperse systems,
which would limit the validity of measurements.

(i) First at all, the scale length of the viscometer (i.e. tube radius,
gap width between coaxial cylinders or cone and plate) must be greater
than about fifty times the particle diameter to avoid phase separation
effects. For blood, this requires a gap larger than about 500 µm. However, this value will be too small at low shear rates since particle aggregates are present, the lower the shear rate, the larger the aggregate
size. That leads to the formation of a marginal zone near the walls. Simultaneously, sedimentation of these aggregates occurs especially at low
shear rates and seems to give larger error in cone-plate viscometers
than in coaxial cylinder ones. Both phase separation and sedimentation
result in time dependent viscosity measurements under steady conditions,
(QUEMADA et al., 1981) then in difficulties in low shear viscometry.

(ii) Interface effects come from the formation of a film of proteins at the blood-air interface. Mechanical properties of this film can perturb the viscosity measurement, especially in the case of the plasma. A guard ring can prevent such an error (BROOKS et al., 1970).

(iii) As the onset of turbulence occurs in particle suspensions at critical Reynolds Number ($\sim$ 10) much smaller than Re_c ($\sim$ 2000) for ordinary fluids, it is necessary to insure that high shear rate measurements remain free from such effects.

3 . BLOOD AS A HIGHLY CONCENTRATED SUSPENSION : EFFECT OF HEMATOCRIT.

3.1. Variables governing rheological properties of blood and RBC suspensions under steady conditions: (1) variations of η with hematocrit at
given shear rate.

     Apparent viscosity of blood, withdrawn with addition of anticoagu-
lant, varies as a function of the sample hematocrit H and the applied
shear rate $\dot{\gamma}_a$. Apparent relative viscosity η_{ra} is :

$$\eta_{ra} = \frac{\eta_a}{\eta_P} = \eta_{ra}(H , \dot{\gamma}_a) \tag{3.1}$$

where η_P is the plasma viscosity. This function verifies the following
relations :

$$\frac{\partial \eta_{ra}}{\partial H} > 0 \quad ; \quad \frac{\partial \eta_{ra}}{\partial \dot{\gamma}_a} < 0 \tag{3.2}$$

that is (i) as for any suspension, blood viscosity increases when volume
fraction grows and (ii) blood is a non-newtonian (shear-thinning) fluid.

     Eq. (3.1) is in agreement with general conclusions of dimensional
analysis (See § 2.1) for local values of the relative viscosity $\eta_r = \eta_r(H , \dot{\gamma}_r)$ under steady conditions. The present section is devoted to
the analysis of concentration effects on blood viscosity at fixed applied
shear rate, $\eta_r = \eta_r(H)$ especially at very low or very high shear rates
when the suspension behaves as a newtonian fluid.

     Fig. 3.1, from CHIEN et al., (1971 a) shows such variations $\eta_{ra}(H)$
for normal human RBC suspended in plasma and in Ringer solution (°). Although these viscosities are much smaller than those of suspensions of

----

(°) which is a saline solution (0.9% (w) of NaCl) containing small amounts

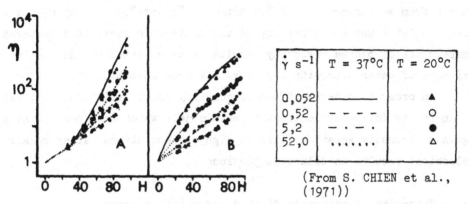

(From S. CHIEN et al., (1971))

Fig. 3.1 Relative apparent viscosity η_{ra} of normal RBC
         suspended in saline (A) or plasma (B) as a func-
         tion of hematocrit H , at different values of
         shear rate $\dot{\gamma}$, and at two temperatures.

rigid particles at same volume fraction (See Fig. 1.5), rapid variations
are observed near the physiological level in hematocrit. Measurements
were performed in a coaxial-cylinder viscometer, at two temperatures
(20°C (points) and 37°C (curves)). These results give proof that the essential part of temperature effects are involved in the plasma viscosity

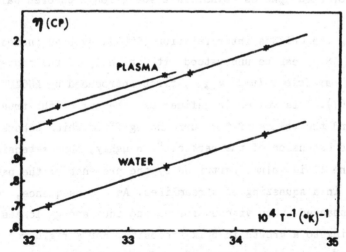

Fig. 3.2 Temperature variation of plasma viscosity (different
         protein content) (from COKELET, 1972).

since points measured at 20°C are very close to the experimental curve
drawn from measurements at 37°C, although "plasma" viscosity of the former was 1.5 times the viscosity of the latter. In fact, this temperature
dependence of plasma viscosity results in main part from the similar variation of water viscosity (See Fig. 3.2 from COKELET, 1972).

In order to understand and to quantify these variations we shall return in the following sections to main models which have been built to
gain a viscosity equation valid at high concentrations, after recalling
classical results on dilute suspensions.

3.2. Viscosity of extremely diluted and dilute suspensions.

In a very dilute system, the suspension viscosity η has been observed higher than the suspending fluid viscosity η_F . The relative viscosity η_r , as a function of the *volume* concentration ϕ , is given by :

$$\eta_r = \frac{\eta}{\eta_F} = 1 + k_1 \phi \qquad\qquad (3.3)$$

where k_1 is a particle shape dependent factor, thus which depends on
particle deformability (through the state of deformation reached in the
experiment).For hard spheres, EINSTEIN (1905), found theoretically $k_1 = k_E = 2.5$.

Following the CHIEN's interpretation (1972), such an increase in
viscosity ($\eta_r > 1$) can be understood with the help of the concept of
the *effective particle volume* v_{eff} , first introduced by ANCZUROWSKI
and MASON (1967). This volume is defined as "the sum of the true particle
volume v_p and the volume of the surrounding fluid which behaves as if
it was a rigid extension of the particle". Roughly, this extension corresponds to the fluid volume perturbed by the presence of the particle,
which results in a squeezing of streamlines. As a consequence, velocity
gradient is enhanced, then viscous forces and thus energy losses are increased, leading to a growth in bulk viscosity. For a rigid rod or disc
this volume v_{eff}^s is higher than v_{eff}^s for a sphere having the same

volume v_p , leading to $k_1 > 2.5$. On the contrary, $k_1 < 2.5$ for a liquid drop, which undergoes deformation and partial alignment, with rotation of the internal fluid (See Fig. 4 in CHIEN's paper, 1972).

As eq. (3.3) contains only *particle-fluid interactions* discarding any hydrodynamic action of one particle upon another, this equation only holds at extremely low concentrations, that has been an important difficulty in experimental verification of the Einstein's result. Above $\phi \sim$ 0.01 - 0.05 , deviation from linearity is observed and attributed to particle-particle interactions. The simplest interaction involves two single particles and give the ϕ^2 term in a power series for relative viscosity :

$$\eta_r = 1 + k_1 \phi + k_2 \phi^2 + \dots \qquad (3.4)$$

A number of theories calculated the ϕ^2 coefficient, k_2 , by different methods, but leading to very different values (k_1 being assumed equal to 2.5 for spheres). Accounting for hydrodynamic interaction between spheres by the method of successive reflections and doublet formation by collisions, VAND (1948) obtained $k_2 = 7.349$. Nevertheless, the most rigorous calculation of shear viscosity recently performed (BATCHELOR and GREEN, 1972) gave $k_2 = 5.2 \pm 0.3$ (this imprecision results from difficulties in numerical evaluation of integrals).

Increasing concentration requires more terms in the series (3.4). No theoretical calculation of coefficients k_i exists and their determination can be only carried out by data fitting (see later).

In the case of deformable particles, as bubbles or liquid droplets suspended in an unmiscible fluid, TAYLOR (1932) gave the expression of k_1 as a function of the ratio $\alpha = \eta_i / \eta_F$, η_i being the internal viscosity of particles

$$k_1 = \frac{k_E \alpha + 1}{\alpha + 1} = \tau k_E \qquad (k_E = 2.5) \qquad (3.5)$$

where $\mathcal{T}$ is the Taylor's factor. With Γ as the drop interfacial tension, eq. (3.5) holds in the limit $\beta = (\eta_F \dot{\gamma} a/\Gamma) \ll 1$, that is if particles undergo very small deformation. From eq. (3.5), Einstein's result is recovered with "rigid" particles suspension $(\alpha \gg 1)$, while gas bubble emulsion $(\alpha \ll 1)$ leads to $k_1 \to 1$.

3.3. Viscosity of concentrated suspensions.

For highly concentrated suspensions, phenomenological approaches have been developped. They lead to (phenomenological) viscosity equations, and using them appears better than power series as (3.4). Indeed such relations $\eta = \eta(\phi)$ must represent the strong increase in viscosity up to infinity, as ϕ tends towards its maximum (packing) value ϕ_M (Fig. 3.3).

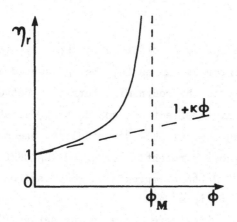

Fig. 3.3 Relative viscosity η_r versus volume fraction ϕ of particles, in concentrated suspensions Infinite viscosity is found at packing concentration ϕ_M .

We shall give a detailled study of these equations and of some purely empirical relations.

3.3.1. Phenomenological equations

A first group of studies refers to "effective medium" theories, considering the suspension with finite concentration ϕ , as a fictive suspending fluid, having a (unknown) viscosity $\eta(\phi)$, to which an incremental fraction of spheres, $d\phi$, is added. The new suspension (concentration $\phi + d\phi$) is then considered as extremely diluted (concentration = $d\phi$) comparing to the first one (considered as the suspending medium). Applying therefore the Einstein's result gives :

$$\eta_r = \eta(\phi + d\phi)/\eta(\phi) = 1 + k_1 \, d\phi \tag{3.6a}$$

that yields the Arrhenius equation (ARRHENIUS, 1917) ;

$$\eta_r = \exp(k_1 \, \phi) \tag{3.6}$$

BRINKMAN (1952) analysed more cautiously the relation between $d\phi$ and the concentration of the (fictive) extremely dilute system. Assuming the latter is obtained by adding one sphere (volume $\boldsymbol{v}_p$) to the N ones contained in a suspension of volume V (thus $\phi = N \boldsymbol{v}_p/V$) , he calculated the concentration increment $d\phi$ from $\phi + d\phi = (N + 1)\boldsymbol{v}_p/(V + \boldsymbol{v}_p)$, that is $d\phi = \boldsymbol{v}_p(1 - \phi)/(V + \boldsymbol{v}_p)$. Applying the Einstein equation (3.6a) to the dilute suspension having the volume concentration $\boldsymbol{v}_p/(V + \boldsymbol{v}_p)$ led to

$$\frac{\eta(\phi + d\phi)}{\eta(\phi)} - 1 = k_1 \frac{d\phi}{1 - \phi} \tag{3.7a}$$

and gave, if $\eta_r(0) = 1$,

$$\eta_r = (1 - \phi)^{-k_1} \tag{3.7}$$

Eq. (3.7a), compared to (3.6a), shows that the particle addition is equivalent to adding the amount $V \cdot d\phi$ to the remaining liquid volume $V(1 - \phi)$, therefore as an excluded volume effect.

By a similar calculation, ROSCOE (1952) obtained (3.7) with an effective concentration ϕ_{eff} which takes into account both particle size distribution and effects of fluid trapping by aggregates. Putting $\phi_{eff} = \lambda \phi$ in (3.7) gives

$$\eta_r = (1 - \lambda \phi)^{-k_1} \tag{3.8}$$

where λ is a concentration dependent empirical factor, close to unity at low concentration, but increasing up to $1.35 = (0.74)^{-1}$ as ϕ

increases and tends to the packing value ϕ_M (close packing of spheres is assumed to be reached here, that is $\phi_M = 0.74$).

It is worthy of note that although using Einstein equation for the fictive dilute system, particle - particle interactions are taken into account through interactions between the added particle and the effective suspending fluid, i.e, through the "effective medium" description.

MOONEY (1951), did not consider, as above, the high dilution limit. He started from a suspension of spheres at ϕ_1 , then having a viscosity $\eta_1 = \eta_F \, f(\phi_1)$ where η_F is the viscosity of suspending fluid and $f(\phi)$ appears as the relative viscosity. This factor must reduce to the Einstein's relation (3.3) as $\phi_1 \to 0$. Adding a finite amount ϕ_2 he obtained a new suspension, the viscosity of which was written $\eta_{1+2} = \eta_1 \, f(\phi_{21})$, where $\phi_{21} = \phi_2/(1 - \lambda \phi_1)$ is the concentration of ϕ_2 in the remaining liquid filling the space not occupied by ϕ_1 , with λ as a crowding factor. Therefore $f(\phi_1 + \phi_2) = f(\phi_1).f(\phi_{21})$. Considering that crowding influences mutually ϕ_1 and ϕ_2 , Mooney similarly took ϕ_{12} instead of ϕ_1 and arrived to the following functional equation for $\eta_r(\phi) \equiv f(\phi)$

$$f(\phi_1 + \phi_2) = f\left(\frac{\phi_2}{1 - \lambda\phi_1}\right) . f\left(\frac{\phi_1}{1 - \lambda\phi_2}\right) \qquad (3.9a)$$

the solution of which is :

$$\eta_r = \exp \frac{k_1 \phi}{1 - \lambda\phi} \qquad (3.9)$$

accounting for the Einstein's limit at $\phi \to 0$. Mooney extended eq. (3.9) to polydisperse suspensions.

KRIEGER and DOUGHERTY (1967) claimed that crowding does not act symmetrically on ϕ_1 and ϕ_2 : depending on the order of operating, one must take ϕ_1 and ϕ_{21} (or ϕ_2 and ϕ_{12}) . Therefore, one must write instead of (3.9a) :

$$f(\phi_1 + \phi_2) = f(\phi_1) . f\left(\frac{\phi_2}{1 - \lambda\phi_1}\right) = f(\phi_2) . f\left(\frac{\phi_1}{1 - \lambda\phi_2}\right)$$

which is satisfied if :

$$\eta_r = (1 - \lambda \phi)^{-k_1/\lambda} \qquad (3.10)$$

where k_1 is again chosen from the dilute limit.

VAND (1948) solved the hydrodynamic equations for fluid flow around a sphere. From the solution, he derived the Arrherius formula (3.6), in the case of a suspension of non-interacting spheres, assuming that the disturbance of streaming caused by the particles changes the original value $\dot{\gamma}$ of the shear rate to $\dot{\gamma} + d\dot{\gamma}$, the shear rate increment $d\dot{\gamma}$ being related to the increase of viscosity $d\eta$ by $d\eta/\eta = - d\dot{\gamma}/\dot{\gamma}$, which results from the conservation of the shear stress $\sigma = \eta\dot{\gamma}$ in a Couette flow. Accounting for hydrodynamic interactions between spheres (calculated by the reflection method) led to change k_1 from (3.6) into a concentration dependent coefficient such as

$$k = k_1 / (1 - Q_v\phi) \qquad (3.11a)$$

where $Q_v = 39/64 = 0.609...$ is an interaction constant. Finally, as collisions between spheres create doublets, triplets, ... a mean shape factor k for the suspension as a mixture of particles of various shapes (shape factors κ_i) at different concentration ϕ_i , was written such as $k \phi = \sum_{i=1}^{\infty} \kappa_i \phi_i$, that led to the viscosity equation (3.6), where k_1 is changed into

$$k = k_1 (1 + a_1 \phi + a_2 \phi^2 + ...) / (1 - Q \phi) \qquad (3.11b)$$

Therefore, η_r resulted in a power series like (3.4). Vand calculated the life time of a doublet and estimated its shape factor (including the effect of the immobilized liquid). Thus he obtained (3.4) for the viscosity of a suspension of spheres, with $k_1 = 2.5$ and $k_2 = 7.349$.

A second group of attempts to $\eta_r = \eta_r(\phi)$ results from application of an energy principle which allows one to derive only general bounds on

viscosity (KELLER et al., 1967). For instance, by the principle of minimum entropy production, S. PRAGER (1963) obtained a lower bound for the
relative fluidity

$$F = \eta_r^{-1} \leqslant (1 - \frac{8}{5}\phi + \frac{3}{5}\phi^2)$$

Nevertheless a new viscosity equation can be obtained from an energy
principle, recalling that, as yet mentioned in §1.3, flows of concentrated suspensions through narrow channels exhibit a two-phase structure.
Taking into account the possible occuring of such a phase separation
with a particle depleted marginal zone near the walls(i.e. for instance
in a pipe, a radial distribution in concentration, $\phi = \phi(r)$), the validity of an optimization principle for viscous dissipation has been postulated (QUEMADA, 1977). As particle motions and flow of suspending
fluid are coupled by hydrodynamic interactions, a *self-consistent* solution - i.e. *both* v(r) and $\phi(r)$ - is believed to exist ,where v(r) is
the velocity profile. As shear stresses induce structural changes, one
assumes in this approach the existence of a (structural) viscosity, $\eta = \eta(\phi,\dot{\gamma})$ will depend not only on properties of the state of the system but
also on flow conditions, the knowledge of which is thus required for de-
fining this viscosity function.

For fixed properties of the whole system (concentration, temperature,
suspending fluid characteristics...) *and* well-defined flow conditions,
such an energy principle states that among all stationary solutions (one-
phase or two-phase viscometric flows), the actual solution (v,ϕ) is the
one which minimizes the rate of energy dissipation (or more generally,
the rate of entropy production). Therefore, any unsteady process is excluded from the present approach, especially mechanisms involved in the
formation of the actual two-phase flow structure.

Considering the flow through a pipe $(0 \leqslant r \leqslant R)$ as an example, minimization of the rate of viscous energy dissipation per unit length of
the tube

$$\mathcal{D} = 2\pi \int_0^R \eta v'^2 r \, dr \qquad\qquad (3.12)$$

(where $v' = dv/dr$) is carried out, accounting for the flow conditions as constraints upon $v(r)$ and $\phi(r)$. Euler-Lagrange's equations result from applying the variational method to a functional which contains Lagrange multipliers associated to the above-mentioned constraints. Assuming, for the sake of simplicity that the system behaves as a newtonian fluid and that the actual concentration profile $\phi(r)$ can be approximated by a rectangular one, leads to decoupling of Euler-Lagrange's equations. This gives (i) the Navier-Stokes'equation for the well-known two-fluid velocity profile and (ii) a functional differential equation for the relative fluidity, $F = \eta_F / \eta$, as a function of volume concentration ϕ . This equation relates values of F and   its derivative $F' = dF/d\phi$ in both the core ($0 \leqslant r \leqslant \beta R$, s subscript) and the marginal layer ($\beta R \leqslant r < R$, w subscript), such as

$$(F'_s + F'_w)(\phi_s - \phi_w) - 2(F_s - F_w) = 0 \qquad (3.13)$$

Eq. (3.13) can be solved in the two following limiting cases which correspond to the one-fluid limits $\beta \to 1$ and $\beta \to 0$ (i) if $\phi_w \to 0$, i.e. for a particle free marginal layer and (ii) if $\phi_s \to \phi_M$, i.e. for a packed core (with zero fluidity).

(i) <u>Particle free marginal layer</u>. Thus, $F_w = 1$ (since $\phi_w = 0$) and (3.13) reduces to

$$F'_s - k_1 + \frac{2}{\phi}(1 - F_s) = 0 \qquad (3.14)$$

where

$$k_1 = \lim_{\phi_w \to 0}(- dF_w/d\phi_w) \qquad (3.15)$$

i.e. appears as the *intrinsic viscosity* yet introduced in (3.4). It can be calculated from theoretical approaches carried out in the high dilution limit (for example, the Einstein's result for neutrally buoyant rigid spheres, $k_1 = k_E = 2.5$). Integration of (3.14) leads to :

$$F = 1 - k_1 \phi + K \phi^2 \qquad (\text{subscript}\ s\ \text{omitted}) \qquad (3.16)$$

where K is an integration constant.

In semi-dilute systems $\phi^2 \ll 1$, the result of OLIVER and WARD (1953) is recovered :

$$\eta_r = (1 - k_1 \phi)^{-1} \tag{3.17}$$

Eq. (3.17) was found in fair agreement with different data, up to $\phi = 0.25 - 0.30$, i.e. in better agreement than (3.3), although these two equations are formally equivalent in the Einstein's limit. Such a difference could be understood through difference in accounting for particle interactions through (3.3) or (3.17). The value of K in (3.16) can be taken from the coefficients k_2 of the ϕ^2 terms in the series (3.4). The Batchelor and Green's k_2 value ; $k_2 = 5.2 \pm 0.3$ leads to K = 1.05 ± 0.3 .

For higher concentration one can determine both k_1 and K from experimental data. Alternatively one can find K = $K(k_1)$ - for a given k_1 value - from a packing concentration limit, $\phi \rightarrow \phi_M$, for which

$$F(\phi_M) = 0 \tag{3.18a}$$

Moreover many systems exhibit a smooth approach to zero of the slope $dF/d\phi$ as $\phi \rightarrow \phi_M$, such as

$$\lim_{\phi \rightarrow \phi_M} (dF/d\phi) \rightarrow 0 \tag{3.18b}$$

Using these conditions (3.18a and b) leads to K = ϕ_M^{-2} and

$$k_1 = 2/\phi_M \tag{3.19}$$

(3.19) appears as a rather surprising result, since it relates the high dilution limit k_1 and the packing concentration. (In fact as we shall see later, k_1 in (3.19) is an effective intrinsic viscosity hereafter named k which involves high concentration effects associated to aggregate shape and suspending fluid trapping. Thus, k will be a ϕ-dependent structural variable, with $k(0) = k_E$ for spheres and $k(\phi_M) = 2/\phi_M$. Eq.(3.19) leads to the very simple form :

$$F = (1 - \frac{1}{2} k \phi)^2 = (1 - \frac{\phi}{\phi_M})^2 \tag{3.20a}$$

that is, for the relative viscosity :

$$\eta_r = (1 - \frac{1}{2} k \phi)^{-2} = (1 - \frac{\phi}{\phi_M})^{-2} \tag{3.20}$$

Notice that (3.20) appears as a special case of the KRIEGER-DOUGHERTY equation (3.10), with $\lambda = k/2$.

In the vicinity of the packing concentration ϕ_M, such expressions call for considering the "solidification" of a suspension, either as a kind of phase transition of liquid-solid type (with a critical exponent 2 for the viscosity) or, may be preferable, as the *sol→gel transition* observed in gelation of colloids.

(ii) Packed Core

However, increasing the feed concentration ϕ and before reaching the above-mentionned sol-gel transition in bulk, a percolation transition could occur if the core concentration ϕ_s reaches the packing value ϕ_M at $\phi = \phi^*$. Nearly above ϕ^*, one has a very thin axial core with $\phi_s = \phi_M$ and a very thick peripheral layer with $\phi_w \simeq \phi$. As ϕ grows, the core thickens at the expense of the peripheral layer. This percolation transition is therefore of the same type than DE GENNES (1979) postulated for suspensions of non-interacting hard spheres. Nevertheless such a transition is believed greatly promoted by attractive interactions between spheres, the higher the latter, the clearer the evidence of this transition. At $\phi \geqslant \phi^*$, $F_s = F_s(\phi_M) = 0$, and if one assumes that $F_s' = 0$ (as in 3.18b), (3.13) becomes

$$F_w' (\phi_M - \phi_w) + 2 F_w = 0 \tag{3.21}$$

Putting $F_w(0) = 1$, the solution of (3.21) is

$$F_w = \left(1 - \frac{\phi_w}{\phi_M}\right)^2 \qquad\qquad \phi \geqslant \phi^* \qquad\qquad (3.22)$$

which gives the fluidity of the suspension in which the core (as an infinite aggregate in the flow direction) is suspended. Assuming that, at $\phi < \phi^*$, (3.17) holds

$$F = 1 - k_1 \phi \qquad\qquad \phi \leqslant \phi^* \qquad\qquad (3.22a)$$

an approximate value of ϕ^* can be obtained by equalling (3.22) and (3.22a) which gives as expected a ϕ_M-dependent value of ϕ^*

$$\phi^* = \phi_M(2 - k_1 \phi_M)$$

For spheres, for instance, one finds $\phi^* = 0.11$ for close packing ($\phi_M = 0.74$) and $\phi^* = 0.26$ for random close packing ($\phi_M = 0.637$). Critical values of ϕ^* between 0.15 and 0.25, have been observed which depend on particle to pipe diameter ratio (KARNIS et al, 1966). Better estimation of ϕ^* could be obtained using (3.16) if K value was more precisely known.

(iii) Discussion

As (3.22) is formally identical to (3.20a) this equation is believed to have some general applicability. Furthermore, these expressions $F_i = (1 - \phi_i/\phi_M)^2$, for $i = s$ or w , verify the general equation (3.13). Indeed, fitting (3.20) on various data using different values of k leads to a satisfactory agreement, as shown on Fig. 3.4.

Nevertheless, more precise fittings give very interesting results. Fig. 3.5 shows the comparison of (3.20) with the experimental data of EILERS (1941) on polydisperse suspensions of asphalt bitumen particles (1.6 μm < 2a < 8.1 μm) and with that of VAND (1948) on nearly monodisperse suspensions of glass spheres (100 μm < 2a < 160 μm). Eq (3.20) is

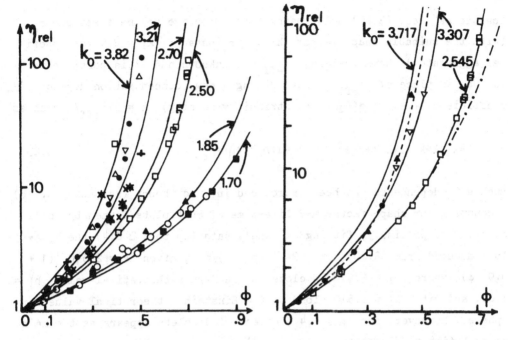

Fig 3.4 Relative viscosity η_r
         versus volume fraction
         ϕ from different data.
         Curves from eq.(3.20)
         with different k_1 va-
         lues.(From QUEMADA, 1977).

Fig 3.5 Comparison between
         Mooney eq(3.9) and
         "Optimum Dissipation"
         eq (3.20) fitting Vand's
         and Eiler's data.

found in very fair agreement with the Eiler's data but not with the Vand's
one . It has been postulated (QUEMADA, 1977) that such a discrepancy
could originate from the difference between the true concentration ϕ
and the effective one ϕ_{eff} , this difference resulting from trapped
fluid by transient doublets, triplets.... In polydisperse systems the
voids between large spheres are filled by smaller ones, that results
in a negligeable fluid trapping, i.e. $\phi_{eff} \simeq \phi$. Then a good fitting is
expected using a constant k_1-value, close to k_E . Morever, close packing concentration can reach values higher than 0.74 (for monodisperse
spheres), up to $\phi_M \simeq 1$ (in the limit of complete polydispersity). On the
contrary, monodisperse systems exhibit ϕ_{eff} larger than ϕ , the ratio
ϕ_{eff}/ϕ depending on ϕ. As the model involves the actual volume of

"particles", eq. (3.20) holds using this effective volume fraction ϕ_{eff} (with the suitable shape factor k_1 : for instance $k_1 = 2.5$ for spherical aggregates). Nevertheless, ϕ_{eff} is unknown and it is possible to use ϕ in place of ϕ_{eff} , introducing a new (concentration dependent) coefficient (i.e. an __effective__ intrinsic viscosity) $k = k_1\phi_{eff}/\phi$ such as

$$k=k(\phi)=k_1(1+b_1\phi+b_2\phi^2+ \ldots) \text{ with } k(\phi_M)= \frac{2}{\phi_M}$$ (3.23)

Such a ϕ-dependence has been introduced in Vand theory (see above 3.11b) accounting for shape factor and life-time of multiplets formed by collisions. More precisely, fitting on Vand's data (up to $\phi \simeq 0.4$) the k values deduced from (3.20), $k = 2(1 - \eta_r^{-1/2})/\phi$, gives $k = 2.510$ (1 + 0.91 ϕ), where $b_1 = 0.91$ is close to the Vand's theoretical value $b_1^V = 1.08$, and with $k_1 = 2.510$ close to the Einstein's theoretical value for spheres. Moreover, for $\phi \leqslant 0.4$, effect of doublets appears as the main contribution to viscosity, since (3.23) reduces to a linear variation.

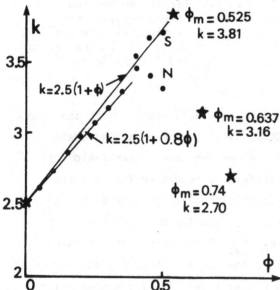

Fig. 3.6 displays these results. The same fitting on Eiler's data gives k = 2.504 (1 + 0.03 ϕ), hence with k_1 very close to k_E and $b_1 \ll 1$. (i.e. negligeable trapping). Similar results have been obtained from data of SAUNDERS (1961), including surfactant layer effects, (QUEMADA, 1978a). On the other hand above $\phi = 0.40$, as shown on Fig. 3.6 the variation $k(\phi)$ changes for stirred (S) or unstirred (N) suspensions. The former gives k-values linearly increasing (i.e. negligeable triplet and multiplet formation under

Fig. 3.6 Intrinsic viscosity vs. concentration, from Vand's data. Packing concentrations ϕ_M and corresponding intrinsic viscosities k = $2/\phi_M$.

stirred conditions) up to the value $k_C = 2/\phi_C = 3.81$ corresponding to
cubic packing ($\phi_C = 0.525$) which is required to allow relative slipping
of spheres. For the latter, k decreases and tends to approach the random
close packing value, ($\phi_{RCP} = 0.625$, $k_{RCP} = 3.20$) or the close packing
one ($\phi_{CP} = 0.74$, $k_{CP} = 2.70$).

Coming back to Fig. 3.5 one can see fittings of the Mooney's equation (3.9) on Eiler's data and Vand's one. They are shown in moderate
agreement, with the reasonable packing value $\lambda^{-1} = 0.70$ for the Vand's
data, but the unacceptable value $\lambda^{-1} = 1.43$ for the Eiler's one. Difficulties in applying the Mooney equation, have been yet stressed
(GILLESPIE, 1963).

The above discussion has shown the importance of the intrinsic viscosity k in (3.20) as a structural parameter, noticely in relation with
the packing fraction ϕ_M . In fact, it must be stressed that in general,
including the case of suspensions of deformable particles, the actual
structure of the suspension at a given ϕ - i.e. actual levels of particle aggregation and deformation - can be described by k(ϕ) or by an actual packing concentration

$$\phi_P = 2/k \qquad\qquad\qquad\qquad\qquad (3.24)$$

which would be reached if particles at ϕ were packed without changing
their actual (aggregated and eventually deformed) state, leading to infinite viscosity. It must be kept in mind that this actual packing ϕ_P
will be usually different from the true one, ϕ_M , owing to structural
modifications which would occur during the (true) rising of the concentration. Therefore, the ϕ-dependent value , $\phi_P = \phi_P(\phi)$ must verify
$\phi_P(\phi_M) = \phi_M$. It appears closely related to the inverse of the "relative
sediment volume", as ROBINSON (1949) introduced.

The previous detailed discussion has been given to stress on
(i) the large applicability of eq. (3.20) as a very simple viscosity
equation,

(ii) the importance of the k coefficient as an "effective intrinsic
viscosity" in which effective particle volume effects are included,
which are under influence of many factors (particle size and shape,
fluid trapping,...), all of them being involved in the effective particle volume v_{eff} ,
(iii) the concentration dependence of k which results from (ii)
and finally
(iv) since v_{eff} depends on flow conditions, the possibility of extending (3.20) to the non-newtonian range through a shear-dependent intrinsic
viscosity $k = k(\dot{\gamma})$. (See later, §4)

3.3.2. Empirical equations

Number of empirical equations $\eta_r = \eta_r(\phi)$ have been proposed. Three
of them which give very good data fitting (and which support eq. (3.20))
will be given here :
(i) Eiler's experiments (1941) were found fitted by the Eiler's equation:

$$\eta_r = \left(1 + \frac{2.5 \phi}{2(1 - b\phi)} \right)^2 \qquad (3.25)$$

with empirical packing value b = 0.78.
(ii) MARON et al (1957), by fitting the Ree-Eyring* (non-newtonian) equation on their data, found :

$$\eta_r = (1 - \kappa \phi)^{-2} \qquad (3.26)$$

(iii) CASSON (1979) recently proposed the relation :

$$\eta_r = \frac{1 - \alpha \phi}{(1 - 1.75 \alpha \phi)^2} \qquad (3.27)$$

---

*See eq. (5.7)

where α is a "material parameter, defined as the ratio of the effective hydrodynamic volume of the particle to its actual volume". Therefore
α appears as the voluminosity first introduced by HOUWINK, (1949) and
it is closely related to k in (3.20). In fact (3.27) does not differ
very much from the following expression, for low and moderate ϕ-values:

$$\eta_r = \frac{(1 - \frac{\alpha}{2}\phi)^2}{(1 - 1.75\,\alpha\,\phi)^2} = \frac{1}{\left(1 - \frac{1.25\,\alpha\,\phi}{1 - \frac{\alpha}{2}\phi}\right)^2} \tag{3.27a}$$

Eq. (3.27a) transforms to the form of (3.20), if one takes

$$k = \frac{2.5\,\alpha}{1 - 0.5\,\alpha\,\phi} \simeq 2.5\,\alpha(1 + 0.5\,\alpha\,\phi + \ldots)$$

Using average value $\alpha = 1.07$ found by Casson, gives $k = 2.68\ (1+0.535\phi)$,
a result in agreement with those given above, noticely, $k(\phi=0) = 2.5\,\alpha$,
i.e. $\alpha = \phi_{eff}/\phi$ as expected. Values close to $k = 2.7$ were observed by
PAPIR et KRIEGER (1970) for suspensions of monodisperse colloidal spheres.
Moreover, the "crowding factor" $(1 - 0.5\,\alpha\,\phi)$ is very close to the one
$(1 - 0.61\,\phi)$ calculated by Vand (1949) from hydrodynamic interactions
between spheres.

3.4. Application to blood and RBC suspensions

The models given above were extensively used to fit blood and RBC
suspension data, obtained from artificially prepared samples*, since normal hematocrit is about 40 - 45 .

As hematocrit is not the true volume concentration ϕ , a difficulty
arises for applying model equations to data. In fact, as it has been
pointed out, hematocrit measurements depend not only on centrifugation

---

*Separating RBC and plasma from centrifugation, removing the "buffy coat"
(platelets and white cells) and mixing remaining RBC and plasma in the
desired proportions.

duration and speed (what can be fixed once for all) but also on particle deformability (and aggregability). Therefore, a "corrected hematocrit" H' as the true volume fraction ϕ has been defined. Under normal conditions, this leads to $\phi = 0.97$ H , but to $\phi = 0.60$ H for hardened cells (CHIEN et al., 1971). This can be understood as follows. Let N be number of particles suspended into the total volume V . Thus $\phi = N v_p/V$, v_p being the particle volume. After packing by centrifugation, the packing fraction is $\phi_M = N v_p/H V$, leading to

$$\phi = H \phi_M \text{ , so } \eta_r = (1 - \frac{\phi_M}{\phi_r} H)^{-2} \tag{3.27}$$

Data are generally given using corrected hematocrit (See Fig. 3.1 for instance).

At moderate concentrations (up to H $\sim$.30) MAUDE and WHITMORE (1958) verified the Oliver and Ward's equation (3.17) with k = 1.7 - 1.9. BAYLISS (1952) found that the same equation (3.17) holds for plasma or serum using k = 0.06 $\pm$ 0.01 $(\text{gr}/100 \text{ ml})^{-1}$.

As hematocrit in pathologic situations varies about from.25 to .70, need of an equation which covers the total range of variations in H led several authors to use equations for highly concentrated suspensions. The more commonly used is the Arrhemius equation (3.6) which leads to a linear dependence of Log η_r vs H . However, Fig. 3.1 shows that such an equation fails at high hematocrit and at low shear rates. Therefore, polynomial fitting has been performed. As an example, CHIEN et al. (1966) gave fifth power polynomial fitting the data shown in Fig. 3.1..

For hardened RBC, the Brinkman-Roscoe equation (3.8) approximatively holds as shown on Fig. 3.7 (from COKELET, 1972). The curve :

$$\eta_r = (1 - \frac{H}{H_M})^{-2.5} \tag{3.28}$$

proposed by LANDEL et al. (1965) using H_M =.635 fits the data of CHIEN et al. (1967) and (1971) on the one hand and the data of BROOKS et al. (1970) on the other hand.

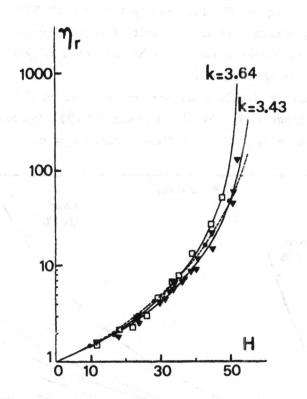

Fig. 3.7 Relative viscosity vs volume concentration
          for hardened cells
                • data of CHIEN et al. (1967)
                ▼ data of BROOKS et al. (1970)
                □ data of CHIEN et al. (1971)
curves ————— eq. (3.20)
       ------- eq. (3.28a with H_M = 0.635 (LANDEL et al. 1965).

DINTENFASS (1968) generalized the Brinkman-Roscoe equation (3.8) to
the form $\eta_r = (1 - \kappa \mathcal{T} \phi)^{-2.5}$, where $\mathcal{T}$ is the Taylor factor, defined
in (3.5), and κ a "packing coefficient or a coefficient of fluid

immobilization". Then assuming $\kappa \phi = H$ (i.e. $\kappa = \phi_M^{-1}$ from (3.27)), he wrote

$$\eta_r = (1 - \mathcal{T} H)^{-2.5} \qquad\qquad (3.29)$$

The validity of eq. (3.29) has been tested by DINTENFASS. However, this proof is only qualitative as it results from superposing calculated and experimental curves given by DINTENFASS (Fig. 3.7 and 3.8 in DINTENFASS, 1968 , as an example).

Coming back to the data for hardened cells shown on Fig. 3.8 , one obtains better fittings using (3.20) than using (3.28). Moreover, different k values are related to the different suspending media ($k \simeq 3.43$

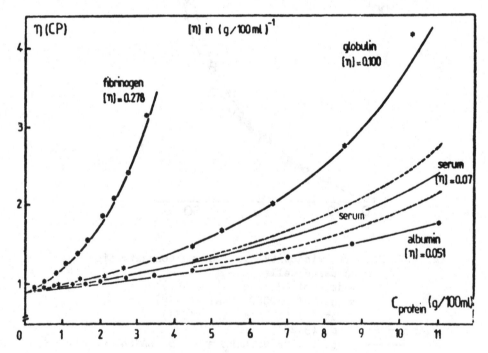

Fig. 3.8 Viscosities of serum protein solutions. Data from
HARKNESS (1971).
Curves $\eta_r = (1 - \frac{1}{2} [\eta]C)^{-2}$ with fitted values
$[\eta]$ (gr/100 ml)$^{-1}$; C = protein concentration (by
weight), (gr/100 ml).

in saline and in Ringer, k ≈ 3.64 in water) that could be related to small changes in the erythrocyte volume due to a slight swelling in water and as a consequence in erythrocyte deformability.

As a viscosity equation valid for any suspension , eq. (3.20) must be also valid for plasma and serum, as suspensions of proteins. Indeed, this validity can be observed using data from HARKNESS (1971), who studied the variations of viscosity of serum and of albumin, globulin and fibrinogen solutions. As volume concentrations of particles are not known, the product $k \phi$ in (3.20) is replaced by the equivalent product $[\eta]C$, using the protein concentration by weight, C , and the (true) intrinsic viscosity, $[\eta]$ measured in $(gr/100 \text{ ml})^{-1}$. Fitting by (3.20) with different $[\eta]$-values are shown in good agreement (Fig. 3.8). For plasma and serum, such values agree with those given by BAYLISS (1952) who found $[\eta] = 0.06 \pm 0.01 \ (gr/100 \text{ ml})^{-1}$, using (3.17) with $2 \leqslant C \leqslant 9$ $(gr/100 \text{ ml})$.

Although such a general agreement would suggest to give more attempts to applying (3.20) to more data, the importance of non-newtonian effects in blood rheology calls for studying now variations of η_r as a function of $\dot{\gamma}$ at constant ϕ , that is the subject of the next section.

4. BLOOD AS A STRUCTURED (SHEAR THINNING) FLUID: (1) Effects of RBC
   Aggregation and Deformation .

In the previous chapter, strong changes in variations of $\eta_r(H)$ when
the shear rate is lowered, has been shown on Fig. 3.1 . Analysis of the
specific effects the RBC Aggregation and RBC Deformation have upon steady
viscosity will be reviewed. Some models for rheological characterization
of blood will be discussed in the next chapter.

4.1. Variables governing rheological properties of blood and RBC suspen-
     sions under steady conditions: (2) Variation of η with shear rate at
     given hematocrit.

Shear thinning behaviour of blood, at constant hematocrit in the
range 40-50%, is a very marked effect, as shown on Fig. 4.1. from BROOKS
et al., 1970. Suspension of normal RBC in saline (NS) exhibits a lower

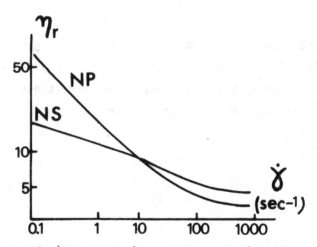

Fig.4.1. Comparison of the relative viscosities of normal red cells suspended
in plasma and isotonic saline. NP : red
cells in plasma H = 53,5%; NS : red cells
in saline H = 52,5%.(From BROOKS et al.,
1971).

viscosity at low shear rate and, on the contrary, a high shear viscosity
higher than the corresponding blood viscosity (NP).

Fundamental determinants of blood viscosity were found by S. CHIEN
(1970). He compared non newtonian behaviour of the following RBC suspensions at same Hematocrit (H=45):
(i) NP = normal RBC in plasma, where fibrinogen and β-globulins promote
the reversible RBC aggregation by molecular bridging to form rouleaux;
(ii) NA = normal RBC in albumin-Ringer solution where no detectable aggregation can be observed;
(iii) HA = hardened RBC (by glutaraldehyde) in albumin-Ringer solution.

The albumin concentration was chosen to give a suspending fluid viscosity (η_F) equal to the plasma viscosity (η_p = 1.2 cP), that is to get
same shear stress at given $\dot{\gamma}$.

Comparing viscosity dependences on shear rate (Fig.4.2) allows to

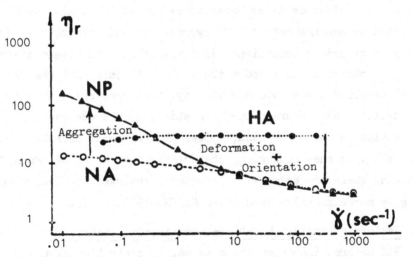

Fig.4.2. Relative apparent viscosity vs. shear rate in three types of suspensions (H = 45%, η_p = 1.2 cP)
NP = normal RBC in plasma, NA = normal RBC in 11% albumin-Ringer, HA = Hardened RBC in 11% albumin-Ringer.
(Adapted from CHIEN, 1970).

separate two fundamental processes in blood rheology:

a) Reversible RBC aggregation at low shear rates, which gives a large increase of NP viscosity in comparison with NA viscosity.

b) RBC deformation and correlative orientation at high shear rates[+] which leads to a same deformation and orientation of single cells in NP (complete dispersion of rouleaux being achieved as in NA), hence to the same reduction in NP and NA viscosities compared to the higher value for HA suspension where such a process obviously cannot occur. Interpretation of RBC deformation-orientation as the process which reduces the high shear viscosity is confirmed by results shown in Fig.4.1:at high shear rates, NP viscosity is lower than NS viscosity because of a lesser RBC deformation in saline than that would be observed in plasma, since the viscosity of the latter ($\eta_F \simeq 0.9$ cP) is lower than the viscosity of the former ($\eta_P = 1.2$ cP).

These two fundamental processes can be studied directly by optical means. Beyond microscopy, methods using light transmission (SCHMID-SCHON-BEIN et al., 1973) or laser backscattering (MILLS et al., 1980) provide quantitative measurements of RBC aggregation and deformation, both under steady or transient conditions, in correlation with stress measurements.

Such observations support the great influence that the "structural state" of blood has on blood viscosity. More precisely, they show that the levels of RBC Deformation-Orientation, on the one hand, and of RBC Aggregation on the other hand, appear as *structural variables* which would allow us the rheological characterization of blood. Main factors governing these processes will be briefly reviewed now following S.CHIEN who gave more exhaustive studies (S.CHIEN, 1972, 1979).

4.2. Factors governing the RBC Deformation-Orientation

RBC Deformation-Orientation is mainly controlled by the following factors:

1) the shear stress $\sigma = \eta_P\dot{\gamma}$ that the plasma exerts upon the RBC. It depends on plasma viscosity η_P and applied shear rate $\dot{\gamma}$. Raising η_P in-

----

(+) nevertheless small enough to avoid any hemolysis effects or the onset of turbulence.

creases RBC deformation, thus RBC alignment in the flow direction, that
leads to a decrease of relative viscosity. This decreasing in η_r has
been observed using RBC suspended in saline, to which was added Dextran
at increasing concentrations (S.CHIEN, 1971). Raising $\dot{\gamma}$ leads to similar
effects, i.e. increases deformation and orientation. These effects hence
constitute the major cause for shear-thinning in the moderately high
shear range (i.e. $\dot{\gamma}$ above about 20 sec^{-1}). In this domain, RBC membrane
exhibits a tank-tread like motion (M.FISCHER et al., 1978), i.e. a membrane rotation around the internal content of the cell. One must notice that
at $\dot{\gamma}$ higher than about 200 sec^{-1}, RBC deformation and alignment are very
close to the limits which were reached if it would be possible to arrive
at $\dot{\gamma} \rightarrow \infty$, avoiding hemolysis to occur.Therefore, a newtonian behaviour
will be recovered at high shear rate (See Fig.5.3). (MERRILL and PELLE-
TIER, 1967).

2) the hematocrit, which promotes RBC deformation by crowding effects
and RBC alignment, the higher the hematocrit the better the alignment
(H.L. GOLDSMITH, 1968), therefore the greater the effective shear rate
between adjacent cells. Thus these effects are the more pronounced as
RBC has a higher degree of deformability. Disappearance of non-newtonian
behaviour at low hematocrit (H $\lesssim$ 10%) would result from the smallness of
crowding and alignment of cells.

3) the pH of plasma, which changes the degree of cell alignment
through modifications of the RBC shape, and deformability (spherocytes
being more deformable than echinocytes). For example in the case of
packed cells (H = 80%) at $\dot{\gamma}$ = 60 sec^{-1}, lowering of the pH from 7.4 to
6.8 gives an increase in viscosity of the order of 60% (DINTENFASS at al.
1966).

4) the intrinsic deformability of RBC, which essentially depends on
the *elasticity of the RBC membrane*, the *viscosity* and the *physico-chemical state of the internal fluid*. The latter could be different from an
ordinary solution (a liquid crystal ?). Abnormalities of hemoglobin molecules as in sickle cells (drepanocytes) containing HbSS, whose rigidi-

ty ($^\circ$) is strongly increased under low pressure in oxygen, leads to very high values in viscosity (DINTENFASS L., 1964; CERNY L.C. and al., 1974). Furthermore RBC deformability was found dependent on its metabolic state (noticely its ATP content, a depletion of which leading to an increase in rigidity). RBC deformability is decreased in the presence of included bodies, as cell nuclei or parasite (MILLER L.H., 1971).

4.3. Factors governing the RBC Aggregation.

Main factors controlling the RBC Aggregation are listed hereafter.

1) The *shear rate* $\dot{\gamma}$ promotes opposite effects:

- on the one hand, *aggregate formation* by "hydrodynamic" collision of two particles, located on adjacent fluid layers. These two particles are in relative motion and then collide, they keep their "contact" over a finite duration of time, allowing aggregation to take place;

- on the other hand, *aggregate rupture* under shear stresses exerted on aggregate by the suspending fluid.

These opposite effects result in a *dynamic equilibrium* between reversible association and reversible dissociation of aggregates. At very low $\dot{\gamma}$, a tri-dimensional (gel-like) network of rouleaux is formed. As $\dot{\gamma}$ is increased, this network is dissociated into single rouleaux, the size of which decreases, that leads to shear-thinning behaviour at low shear. (If RBC aggregability is sufficiently high one can observe rouleau orientation by shear, giving an additional decrease in viscosity). At about $\dot{\gamma} \sim 50 \ sec^{-1}$ (i.e. about $\sigma = 2 \ dynes/cm^2$), all rouleaux are completely dispersed to single cells, the deformation of which is not very far from its very high shear limit.

2) Physico-chemical properties of suspending fluid, play a very important role through the ionic and macromolecular contents of this fluid. Indeed, rouleau formation at low $\dot{\gamma}$ results from competition between cell attraction due to macromolecular bridging, the ends of the macromolecular chain being adsorbed on cell surfaces, and cell repulsion from electrostatic forces. (Remember the lack in RBC Aggregation when RBC are suspended

----

($^\circ$) SS-Hemoglobin is more viscous than the normal one, AA-Hb; that results from Hydrogen bonding and chain cyclization.

in Albumin-Ringer solution which does not contain fibrinogen, see Fig.
4.2, NA). Therefore, RBC aggregation will depend on the chain length
(that is for linear polymers closely related to the molecular weight M)
and on the ionic strength. Substances, as drugs, which modify equilibrium
conditions between these two opposite forces, will act as aggregative or
anti-aggregative substances. For example, Fig. 4.3. shows the decrease in

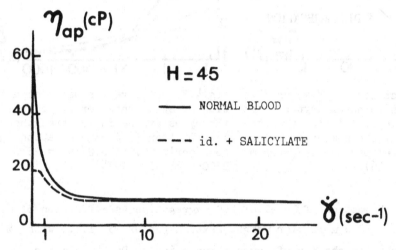

Fig.4.3. Plot of the apparent viscosity vs apparent mean shear rate in stationary flow for normal and unaggregated blood. (From HEALY and JOLY
1975).

low shear viscosity for blood + Aspirin (J.C. HEALY et M. JOLY, 1975).

Evidence of this macromolecular bridging can be found in:

(i) rouleau formation in the presence of Dextran of different molecular weight, which shows a constant interparticle distance d (between
adjacent cells) — thence which requires a cell deformation (mainly of the
concave part) — the distance d depending on M through the length of the
chain (S. CHIEN et al., 1971 c).

(ii) Adsorption isotherms of Dextrans (BROOKS, 1976), which shows an
abrupt change in slope when aggregation starts (Fig. 4.4.), i.e. a factor
2 in slope.

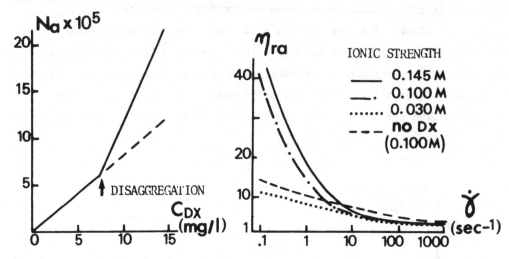

Fig. 4.4. Isotherm of dextran 77.6 adsorption to human erythrocytes at physiological ionic strength in terms of the number of molecules adsorbed per cell, N_a (From BROOKS , 1976).

Fig. 4.5. Relative viscosity as a function of shear rate for washed human erythrocytes at the ionic strengths indicated. Dextran T 70 concentration 3g/100ml; hematocrit 50.0 ± 0.2%; T = 25°C. (From BROOKS et al., 1974).

(iii) Variation of level of RBC aggregation as a function of molecular weight M: Low molecular weight Dextran (as Dx 25, i.e., with M ≈ 25.000) exhibits absence of aggregation owing to Dx 25 is a too short molecule to overpass electrostatic repulsion. The contrary is observed using high molecular weight dextrans (as Dx 150).

(iv) Variation of level of RBC aggregation as a function of ionic strength (BROOKS et al., 1974): the latter governs the spatial extent of electrostatic forces through the ionic double layer thickness, $(\kappa_D)^{-1}$ the higher the ionic strength, the thinner the ionic layer (see Fig. 4.5.).

3) The hematocrit H influences the state of aggregation through the collision frequency f of RBC. At low hematocrit, f is weak and aggregation effects are negligible : newtonian behaviour is recovered at H below about 10% (See. Fig. 5.3.). On the contrary, at high hematocrit, f is too high and impedes the formation of rouleau and blood more and more ressembles to a suspension of non aggregable particles. This leads to

consider RBC aggregation as having its greater effects in the physiological range of hematocrit .

4) The intrinsic aggregability of RBC depends on (see the above discussion):

- RBC membrane affinity for macromolecular adsorption
- Electric charge of the RBC membrane
- RBC deformability, since rouleau formation requires RBC deformation.

In order to progress in quantitative interpretation of the rheological behaviour in terms of structural characteristics of the materials, we shall give now a brief survey of main models proposed for non-newtonian behaviour.

5. BLOOD AS A STRUCTURED (SHEAR THINNING) FLUID: (2) MODELS AND APPLI-CATIONS.

Inspection of fundamental determinants of blood rheology, carried out in the previous section, calls for models as attempts for quantitative evaluation of blood viscosity at different shear rates and under various conditions. As far as possible, such models must involve these fundamental determinants as structural parameters.

Before giving some examples with application to blood rheology, some general models for non-newtonian behaviour will be reviewed.

5.1. Simple rheological models for viscosity of non-newtonian fluids.

For blood, SCOTT-BLAIR (1959) proposed to use the Casson relation which was developped for pigment-oil suspensions (CASSON, 1959). This relation, which gives the shear stress as a function of $\dot{\gamma}$, reads:

$$\sigma^{\frac{1}{2}} = \sigma_0^{\frac{1}{2}} + (K\dot{\gamma})^{\frac{1}{2}} \tag{5.1}$$

where σ_0 and K are "material constants", which depend on concentration; σ_0 is the yield stress, K, the Casson viscosity. In the Casson theory, pigment particles forming long chains by mutual attraction are for the sake of simplicity treated as long rod-like aggregates, which are broken down by shear stresses. At given shear rate $\dot{\gamma}$, a dynamic equilibrium is reached. Casson found semi-empirically an equilibrium value for the aggregate size, in terms of its axial ratio J depending both on $\dot{\gamma}$ and on the cohesive force F between particles in aggregate. The maximum tension on an aggregate compatible with no-rupture must be proportional to $\dot{\gamma}J^2$ leading to $J \sim \dot{\gamma}^{-\frac{1}{2}}$ for a given F. Then he postulated the equilibrium J value given by a linear relation $J = \beta_1 + \beta (\eta_F\dot{\gamma})^{-\frac{1}{2}}$. Then (i) adding energy dissipation in fluid and solid parts of the suspension and (ii) extending the result to high concentrations, he obtained eq. (5.1), with:

$$K = \eta_F(1 - \phi)^{-q} \tag{5.2}$$

$$\sigma_0 = (\frac{a\beta}{q})^2[(1 - \phi)^{-\frac{q}{2}} - 1]^2 \tag{5.3}$$

with $q = (a\beta_1 - 1)$, where a, β_1 and β are structural parameters.

Number of phenomenological or empirical relations $\eta(\dot\gamma)$ have been proposed for general description of pseudo-plastic behaviour. Relations which contain two parameters, as the well known power law $\eta = A \, \dot\gamma^n$, can represent only a part of the involved range of $\dot\gamma$ (Such relations will be not considered hereafter). On the contrary, using three or more parameters allows both low and high shear behaviour to be displayed.

Three parameter relations can be written in the general form:

$$\eta = \eta_\infty + \frac{\eta_0 - \eta_\infty}{F(\dot\gamma)} \qquad\qquad (5.4)$$

where η_0 and η_∞ are the limiting viscosities as $\dot\gamma \to 0$ and $\dot\gamma \to \infty$ respectively. $F(\dot\gamma)$ is a dimensionless function of shear rate such as $F(0)=1$ and $F(\infty) = \infty$. Table 5.1. gives some examples of $F(\dot\gamma)$.

TABLE 5.1

Shear-dependence of $F(\dot\gamma)$

Ref	Remarks	$F(\dot\gamma)$	
WILLAMSON (1929)	empirical, A = Const.	$1 + A\dot\gamma$	(5.5)
CROSS (1965)	semi-empirical, m = $\frac{2}{3}, \frac{4}{5}$	$1 + A\dot\gamma^m$	(5.6)
REE-EYRING (1955)	theoretical (with 2 components, τ = relaxation time	$\tau\dot\gamma / \operatorname{Sinh}^{-1}(\tau\dot\gamma)$	(5.7)

For suspensions of colloidal spheres (radius = a) KRIEGER AND DOUGHERTY (1959) gave a relation similar to eq. (5.4) :

$$\eta = \eta_1 + \frac{\eta_2 - \eta_1}{1 + b\sigma_r} \qquad\qquad (5.8)$$

where : $\sigma_r = \sigma a^3 / K T$ (5.8a)

is a reduced (dimensionless) shear stress, defined from the actual shear stress σ . In (5.8), η_1 and η_2 are the viscosities at the higher and lower shear stresses available respectively. These values are both dependent upon the volume concentration ϕ in the form given by eq. (3.10). In their

theory, Krieger and Dougherty applied reaction kinetics to doublet formation. Nevertheless, they supposed that the concentration of doublets is
very small in comparison with the single particle one, assumption which,
strictly speaking, limits the use of eq. (5.8) to dilute systems.

5.2. Extension of the newtonian model deduced from optimization of energy dissipation to the description of non-newtonian behaviour. (QUEMADA, 1978 a).

Attempt to extend to the non-newtonian range the equation (3.20)
originated from the concept of structural intrinsic viscosity $k = k(\phi)$,
yet introduced in § 3.3. Reversible aggregation of particles was considered as responsible of such a ϕ dependence. Moreover, the latter is believed to result from other "structural" effects (in a large sense), as
thickness of interfacial layers (ionic double layer, surfactant layer)
and particle orientation, the greater the particle deformability, the more
orientated the particles. As most of these effects are shear dependent,
k must be also shear-dependent. Therefore, as an approximation, it will be
supposed that validity of eq. (3.20) can be extended from newtonian to
non-newtonian range , assuming that these dependences, $k = k(\phi,\dot{\gamma})$ reflect
all non-newtonian effets. As resulting from dimensional analysis (§2.1)
this extension necessarily gives $k = k(\phi,\dot{\gamma}_r)$, where the reduced shear rate $\dot{\gamma}_r = \tau\dot{\gamma}$ involves some characteristic time τ. That leads to

$$\eta_r = \eta_r(\phi,\dot{\gamma}_r) = [\, 1 - \frac{\phi}{2} \, k(\phi,\dot{\gamma}_r)]^{-2} \qquad (5.9)$$

Eq.(5.9) is a non-newtonian viscosity equation for very highly concentrated systems. Although (3.20) has been only established in newtonian range, such an extension is believed acceptable,since at very high ϕ and
moderate flow rate (°), the two-phase flow model used in § 3.3. can be
expected still valid, with a quasi-rigid core (where $\dot{\gamma} \ll \tau^{-1}$) surroun-

----

(°) very low and very high flow rates lead to newtonian behaviour

ded by a very thin marginal layer (where $\dot{\gamma} \gg \tau^{-1}$), therefore leading to viscosities close to the (newtonian) low and high shear limits, respectively (See Fig.1.3).

The validity of (5.9) can be tested using various data. As an example for RBC suspended in saline, variations of η_r vs.H at different given shear rates are shown on Fig.5.1. They can be represented by (3.20) with cons-

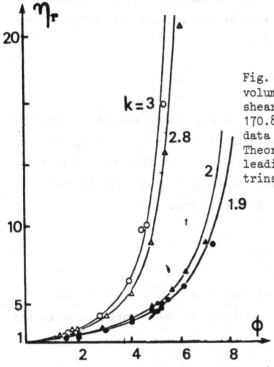

Fig. 5.1. Relative viscosity η, vs volume concentration ϕ at different shear rates $\dot{\gamma}$ [sec^{-1}]= 678.2 ($\bullet$), 170.8 ($\blacktriangle$), 1.708 ($\triangle$), 0.678 (O) (from data of BROOKS and SEAMAN, 1971). Theoretical variations $\eta_r=(1-\frac{1}{2}k\phi)^{-2}$, leading to a shear rate dependent intrinsic viscosity $k = k(\dot{\gamma})$.

tant k, the value of which varies with $\dot{\gamma}$ but not with H since aggregation does not occurs in saline.

At given concentration, the intrinsic viscosity $k = k(\dot{\gamma}_r)$ as a structural parameter will reach limiting values:

$$k_o = k(o) \text{ and } k_\infty = k(\infty) \qquad (5.10)$$

These values can be approximatively obtained using $\tau\dot{\gamma} \ll 1$ and $\tau\dot{\gamma} \gg 1$ respectively, under steady conditions. In fact, if a uniform angular velocity is suddenly applied to a shear-thinning system, this structural parameter k becomes time-dependent, until, a steady state, characterized by

(5.9) will be reached. As an exact time dependent description is not
available, a rate equation can be written that expresses the balance between aggregate build-up due to collisions and aggregate break-down induced by shear. It is assumed that such two opposite processes can be
roughly described by shear dependent characteristic (relaxation) times
τ_A and τ_D respectively. For instance, if one assumes that the system
contains N "particles" of two kinds, aggregates and single (dispersed)
Particles, a rate equation which describes the dynamic equilibrium between these two populations, can be written

$$\frac{dn}{dt} A = \tau_A^{-1} (N - n_A) - \tau_D^{-1} n_A$$

where n_A and $n_D = N - n_A$ are the number of aggregates and single particles, respectively. One can define k in (5.9) by $k\phi = k_A\phi_A + k_D\phi_D$ using
the intrinsic viscosities k_A of aggregate and k_D of single particle suspensions and the corresponding volume fractions $\phi_i = n_i v_p/V$, (i = A,D),
where v_p = single particle volume. At equilibrium

$$n_A = N (1 + \theta)^{-1}, \text{ with } \theta = \tau_A/\tau_D.$$

Writing $k_A \equiv k_o = k(0)$ = intrinsic viscosity at zero shear rate and
$k_D \equiv k_\infty = k(\infty)$ = intrinsic viscosity at infinite shear rate, leads to the
shear dependent intrinsic viscosity

$$k = k_\infty + \frac{k_o - k_\infty}{1 + \theta}$$

where $\theta = \theta(\dot{\gamma})$ with $\theta(0) = 0$ and $\theta(\infty) \to \infty$.

More generally, in order to avoid a too complicated description,
one can postulate the existence of a balance equation describing the
time evolution of the structure   , as resulting from a dynamic equilibrium between its building up and its breaking down, in a large sense,
taking these (reversible) processes in a large meaning (such as

orientation $\rightleftarrows$ disorientation, deformation $\rightleftarrows$ undeformation, aggregation $\rightleftarrows$ disaggregation). Using k as a structural variable leads to the following balance equation:

$$\frac{dk}{dt} = \tau_A^{-1} (k_o - k) - \tau_D^{-1} (k - k_\infty) \qquad (5.11)$$

where τ_A and τ_D are the relaxation times for formation and dissociation of the structure.

Assuming the existence of such mean relaxation times, the steady solution of (5.11) is again

$$k = k_\infty + \frac{k_o - k_\infty}{1 + \theta} \qquad (5.12)$$

where:

$$\theta = \tau_A / \tau_D \qquad (5.13)$$

In the case of dilute systems of rigid spheres (radius a), τ_A^{-1} can be taken as the Browman collision frequency (SMOLUCHOWSKI, 1917) and τ_D^{-1} as the frequency of shear induced collisions (GOLDSMITH and MASON, 1967), that leads to:

$$\theta = \tau_D^{-1} / \tau_A^{-1} = \frac{8\phi\dot{\gamma}}{\pi} / \frac{2\phi}{\pi} \frac{KT}{n_F a^3} = \frac{4 n_F a^3}{KT} \dot{\gamma} \qquad (5.13a)$$

Eq. (5.13a) exhibits a mean characteristic time τ such as $\theta \sim \tau\dot{\gamma}$, in aggrement with the result of dimensional analysis (.28), i.e. $\theta \sim \dot{\gamma}/D_{rot}$.

For concentrated suspensions, the shear dependence of $\theta = \theta(\dot{\gamma})$ is unknown. However, according to the high and low shear limits $\theta(\infty) \to \infty$ and $\theta(0) \to o$, one can take approximatively the following form:

$$\theta = (\tau\dot{\gamma})^p \equiv (\dot{\gamma}/\dot{\gamma}_c)^p \equiv \dot{\gamma}_r^p \tag{5.14}$$

where τ = a characteristic time of the structure and p an empirical parameter. One may define a critical shear rate $\dot{\gamma}_c = \tau^{-1}$, when the order of magnitude of the corresponding value of the shear energy — as the work done by shear forces, $\sigma_c a^2 = \eta_F \dot{\gamma}_c a^2$ on a distance comparable to the particle radius — becomes close to the thermal energy:

$$\eta_F \dot{\gamma}_c a^3 \sim KT \tag{5.14a}$$

in agreement with dimensional analysis (See § 2.1)

Eq. (5.9), through (5.14), involves only one relaxation time τ, what could be a too rough approximation since, in most cases, description of complex systems requires several or sometimes a distribution of relaxation times. If the relaxation time spectrum is composed of several "groups" which can be separated into different "flow units", each group being characterized by a mean relaxation time τ_i it will be possible to use the general model of REE-EYRING (1955) in the form (See eq.(5.7) table 5.1):

$$\eta = \eta_\infty + \sum_i a_i \frac{\tau_i \dot{\gamma}}{\text{Sinh}^{-1} \tau_i \dot{\gamma}} \tag{5.15}$$

Nevertheless, a satisfactory description of the non-newtonian behaviour is very often obtained using only one relaxation time (especially if the latter is the only one whose reciprocal value lies in the shear rate interval accessible to the measuring apparatus). Therefore, unless if

necessary (°) the simple form (5.9 - 5.12) will be kept for the non-newtonian viscosity:

$$\eta_r = (1 - \frac{1}{2} k\phi)^{-2} , \quad k = k_\infty + \frac{k_0 - k_\infty}{1 + (\tau\dot\gamma)^p}$$

(5.16)

Using viscosity values at zero and infinite shear, η_0 and η_∞, respectively, transforms (5.16) to:

$$\sqrt\eta = \sqrt{\eta_\infty} + \frac{\sqrt{\eta_0} - \sqrt{\eta_\infty}}{1 + A(\tau\dot\gamma)^p} \quad \text{where } A = \sqrt{\frac{\eta_0}{\eta_\infty}}$$

(5.16a)

Finally, it must be stressed that instead of a shear rate dependent
equation, the same model would lead to a shear stress dependent viscosity, as in (5.8). Indeed, in terms of effective medium, effects of particles interactions, occuring at high concentrations, can be taken into
account by using in (5.13a), the viscosity of the suspension as a whole
instead of the suspending fluid viscosity. This changes θ , given in
(5.14) to a reduced shear stress, $\theta \sim (\sigma_r)$ where σ_r is defined as in
(5.8a). Put into (5.12), this new θ transforms (5.9) to the simple relation:

$$\sqrt\eta = \sqrt{\eta_\infty} + \frac{\sqrt{\eta_0} - \sqrt{\eta_\infty}}{1 + A(\sigma_r)^p} \quad \text{with } \sigma_r = \sigma/\sigma_c , \quad \sigma_c = KT/a^3$$

(5.17)

very similar to (5.8).

Both (5.16a) and (5.17) are related to pseudo-plastic behaviour, as
shown generally by (5.4). However, an important feature of the present
model, eq. (5.9), is the existence of an actual packing concentration,
defined as in (3.24), now both ϕ and $\dot\gamma$ dependent. At $\dot\gamma = 0$, for instance,

----

(°) For instance, if aggregation of particles induced by shear becomes important, it would be necessary to introduced a new characteristic time τ'
into the rate constant for association, for instance $\tau_A^{-1} = \tau_{A0}^{-1}[1 + (\tau'\dot\gamma)^s]$.
with some new exponent s and where $\tau_{A0} = \tau_A(\dot\gamma = 0)$.

there exists a packing concentration ϕ_{Mo} If one increases ϕ up to ϕ_{Mo} the
zero shear viscosity η_0 tends to infinite, and the pseudo-plastic behaviour
changes to a true plastic one, at $\phi \geqslant \phi_{Mo}$ in the special case $\eta = 1/2$ (see
Chap. 7, and **Fig.** 5.6 and 5.7).

5.3. Application to blood and RBC suspensions

Most of analysis of non-newtonian data for blood and RBC suspensions
used the Casson equation (5.1). MERRILL et al. (1965) found that this
equation correlates very well data for normal blood, but not for washed
red cells suspended in fibrinogen free plasma. Fig.5.2 shows variations
of $\sigma^{\frac{1}{2}}$ versus $\dot{\gamma}^{\frac{1}{2}}$ (Casson plot).

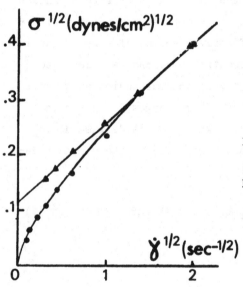

Fig.5.2 $\sigma^{\frac{1}{2}}$ vs $\dot{\gamma}^{\frac{1}{2}}$ for whole blood ▲
(H=41.9) and RBC suspended in de-
fibrinated plasma ● (H=40.6).(From
MERRILL et al. 1965).

Moreover Fig.5.3 from BROOKS et al (1970), shows same variations for
normal blood at different hematocrits. Fair agreement is observed, exclu-
ding parts of data corresponding to very low and very high shear rates,
more especially as H is large (for instance $1 \lesssim \dot{\gamma}^{\frac{1}{2}} \lesssim 20$ sec$^{-\frac{1}{2}}$ for normal
hematocrit). Pseudo-plastic behaviour results from no measurable yield

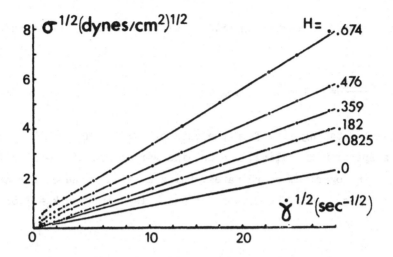

Fig. 5.3 Casson plot $\sigma^{\frac{1}{2}}=f(\dot{\gamma}^{\frac{1}{2}})$ for a blood with ACD anticoagulant. Data of BROOKS et al (1970) at different hematocrit. (Data points for plasma and H = 8.25 RBC suspension are not shown for clarity: these fluids appear to be newtonian. Casson law (4.1) is satisfied except close to $\dot{\gamma}$ = 0, where pseudoplastic behaviour is observed (From G.R. COKELET, 1972).

stress at low shear rate. At high shear rate, transition towards newtonian behaviour (which slightly appears on Fig. 5.3 for H = 67.4) has been observed (MERRILL et al., 1967).

Few other attempts have been performed, as the use of an empirical shear dependent Taylor coefficient inserted in the Brinkman-Roscoe equation (3.8) (DINTENFASS, 1971). However, as this shear dependence is assumed to result only from a shear dependent internal viscosity of the cell, such an attempt is very questionable.

Applying eq. (5.16) first needs to know the value of p, the empirical determination of which can be obtained as follows. From data, $\eta_r = \eta_r(\dot{\gamma})$ at given hematocrit, the effective intrinsic viscosity can be calculated as:

$$k = k(\dot{\gamma}) = 2 \, (1 - \eta_r^{-\frac{1}{2}})/H \qquad\qquad (5.18)$$

and then it can be inserted in (5.12), written as a linear function of log $\dot{\gamma}$:

$$\log \frac{k_0 - k}{k - k_\infty} = p \log \dot{\gamma} + p \log \tau \qquad (5.19)$$

As an example, taking $k(\dot{\gamma})$ from the data of BROOKS and SEAMAN (1971) for RBC suspension in saline, one obtains variations of $Z = (k_0-k)/(k-k_\infty)$ against $\dot{\gamma}$, in logarithm coordinates, using $k_\infty = 1.80$ and different k_0 values (See Fig. 5.4). The linear variation is seen when $k_0 = 3.60$ with a

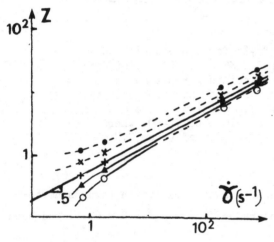

Fig.5.4 Graphical representation of eq. (5.19), using $k_\infty=$ 1.80 for different values of k_0: 3.24 (o), 3.42 (Δ), 3.60 (+) 3.96 (x), 4.50 ($\bullet$).(From QUEMADA 1978 a)

slope p close to 0.5 . More precisely, the best fit linear regression eq.(5.19) is found for the following values:

$$k_0 = 3.69 \quad , \quad k_\infty = 1.78 \quad , \quad \tau = 0.43 \text{ sec}, \quad p = 0.47$$

Number of such comparisons, especially for normal blood data, yielded different values for k_0 , k_∞ and τ but $0.4 \lesssim p \lesssim 0.6$. Thus for simplicity, the value p = 0.5 has been chosen. On the contrary, since p values very close to 1 have been found for uniform colloidal spheres (QUEMADA, 1978 b), values $p \neq 1$ where believed,in suspensions of non-spherical particles, as resulting from polydispersity or/and particle shape. Indeed , in such systems, the non-linearity in $\dot{\gamma}$ dependence might result from variable angular velocities of particles, which spend more time aligned with the flow than normal to it. (Alignment of particles means alignment of faces for simple RBC as a disc, or alignment of axis for a rouleau as a rod; notice that suspended discs and rods exhibit same period of rotation under shear if their equivalent axis ratios are reciprocal (GOLDSMITH and MASON, 1967)).Furthermore, as we will see later (§ 8), such a non linearity could also originate from the smaller value of the shear rate in the core (where blunting of the velocity profile occurs) than the value of the applied shear rate (that is the value which enters in data fitting equations).

As an example, result from fitting (5.16), with $p = \frac{1}{2}$, on the data of MERRILL et al. (1965), is shown on Fig.5.5 (See Fig.5.2 for comparison). It must be stressed that the lack of any yield stress for blood

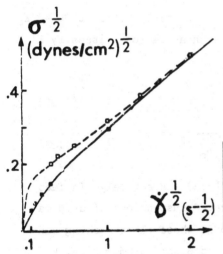

Fig.5.5 Non newtonian behaviour of blood (□) and red cells in defibrinated plasma (■): Data from MERRILL et al., 1965. (See Fig.5.2). Theoretical variations, according to eq. (5.17) with the fitted parameters.

	ϕ	k_∞	k_0	τ	
Blood	0.419	1.80	4.68	1.45	---
RBC in Defibr. Pl.	0.406	1.80	3.96	0.01	——

does not mean that some true yield stress would not exist, but that if it existed, it would be very smaller than the σ_0 value given by the Casson plot.

The Casson equation can be recovered from (5.16) with $p = \frac{1}{2}$ in two limiting cases (QUEMADA, 1978 b):

a) In the "high shear rate" limit, $\tau\dot{\gamma} \gg 1$, (See Fig.5.6),if volume

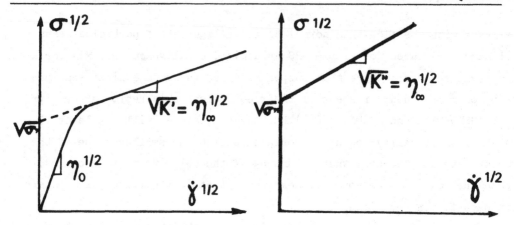

Fig.5.6 Recovering of the Casson equation in the high shear rate limit of (5.16), $\tau\dot{\gamma} \gg 1$ (with a *pseudo-yield* shear stress, σ_0').

Fig.5.7 Recovering of the Casson equation in the very high concentration limit of (5.16), $\phi \to \phi_{Mo}$ (with a *true* yield shear stress, σ_0'').

fraction ϕ is less than the packing value for particles at rest, $\phi_{Mo} = 2/k_o$ Eq. (5.16) then turns into (5.1), written with a "Casson" viscosity K' and *a pseudo-yield stress* σ_0' such as :

$$K' = \eta_F \left(1 - \frac{1}{2} k_\infty \phi\right)^{-2} \equiv \eta_\infty \tag{5.20}$$

$$\sigma_0' = \frac{1}{4} \frac{\eta_\infty^2}{\eta_F \tau} (k_0 - k_\infty)^2 \, \phi^2 = \frac{\eta_\infty}{\tau} \left(1 - \sqrt{\frac{\eta_\infty}{\eta_0}}\right)^2 \tag{5.21}$$

which are closely related to eq. (5.2) and (5.3) respectively. In fact both eq. (5.2) and (5.3) apply to concentrated suspensions of deformable particles, like RBC suspensions or Blood, whose close packing concentration of *disperse* particles $\phi_{M\infty} = 2/k_\infty$ is close to unity, i.e. $k_\infty \approx 2$. Moreover, MERRILL et al. (1965 a) found the exponent q in (5.2) and (5.3) close to q = 2 (average value $\bar{q} = 1.97$), which agrees to the exponent value in (5.20) but that leads to $\sigma_0' \sim \phi^2(1-\phi)^{-2}$ instead of $\sigma_0' \sim \phi^2(1-\phi)^{-4}$ in (5.21).

   b) The pseudo-plastic behaviour can change into a plastic behaviour if at rest the suspension form a very loose network of aggregates characterized by a relatively low "packing" value ϕ_{Mo}. Then, increasing the volume fraction, a *"high concentration" limit* $\phi \to \phi_{Mo} = 2/k_o$ can be rea-

ched. In this limit, the medium at rest forms a three dimensional network
like a (physical) gel, for which $n_0 \to \infty$ (Fig. 5.7.),which still leads to
(5.1) with the same Casson viscosity (5.20), i.e.:

$$K'' = n_\infty = n_F \ (1 - \frac{k_\infty}{k_0})^{-2} \tag{5.22}$$

and a *true-yield-stress* σ_0'' given from (5.11), by:

$$\tilde{\sigma_0} = n_\infty / \ \tau = n_\infty \dot{\gamma}_c \tag{5.23}$$

As evidence σ_0'' is the shear stress required to disrupt the network structure (as in a shear induced gel $\to$ sol transition). Since the presence of
a yield stress is associated to the existence of a partial plug flow in
a tube, this high concentration limit is believed to be reached inside
the core of a two phase structure. Therefore, the present model relates
the plastic behaviour to the packing transition. It has been suggested
(QUEMADA, 1981) that the same model associates the onset of non-newtonian
effects (i.e. pseudo-plastic behaviour) to the percolation threshold introduced in §3.3. . We shall return on this discussion in Chapter 7.

At $\dot{\gamma} = \dot{\gamma}_c$, the shear stress takes the critical value σ_c, such as:

$$\sigma_c = n_c \dot{\gamma}_c = (\sqrt{\tilde{\sigma_0}} + \sqrt{n_\infty \dot{\gamma}_c})^2 = 4\tilde{\sigma_0} \tag{5.23a}$$

This result leads to a very simple graphical method for $\dot{\gamma}_c$ determination
from Casson plot (See Fig. 5.7).

The relations (5.23) and (5.23a) hold approximately in the case of
pseudo-plastic behaviour, provided that $n_0 \gg n_\infty$.

Interpretation of k_0, k_∞ and τ as parameters for rheological characterization of blood can be deduced from analysing the data of CHIEN described in § 4.1. (See.Fig.4.2). Data fitting of eq. (5.16),with $p = \frac{1}{2}$,
leads to values given in table 5.2 .

TABLE 5.2 : Rheological parameters for non-newtonian behaviour.

	$\phi = 0.45$		$\eta_P = 1.2$ cP	
	k_∞	k_0	τ_{sec}	ϕ_{Mo}
NP	1.78	4.20	0.2	0.476
NA	1.78	3.29	0.04	0.608
HA	3.62	–	–	0.552

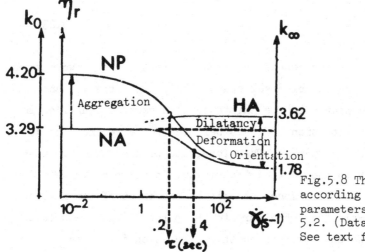

Fig.5.8 Theoretical curves according to rheological parameters given in Table 5.2. (Data from CHIEN, 1970) See text for discussion.

Fig.5.8 shows the corresponding calculated curves. Such values suggest the following interpretations.

(a) k_∞ reflects RBC deformation and orientation, since (i) NP and NA suspensions at $\tau\dot\gamma \gg 1$ contain only single cells which experience same deformation and orientation (because ϕ and η_P have same values in both cases) and (ii) these effects are quite absent from those in HA suspension thus leading to $k_\infty^{NP} \approx k_\infty^{NA} < k_\infty^{HA}$.

(b) k_0^{NP} reflects RBC aggregation since, at $\tau\dot\gamma \ll 1$, (i) neither shear deformation nor orientation of cells occur in NP or NA suspensions, and (ii) aggregation effects appear in NP in comparison with NA, as the latter from which plasma proteins (fibrinogen and globulins) are absent does not undergo cell aggregation, that leads to $k_o^{NA} < k_o^{NP}$.

(c) Packing concentration, as $\phi_{Mo} = 2/k_o$ for the NP suspension $\phi_{Mo}^{NA} = 0.608$, very close to the one, $\phi_o = 0.61 \pm 0.01$ obtained for centrifugal packing of cells hardened in acetaldehyde (CHIEN et al., 1971). As deformation of normal cells by crowding can be expected small enough to be of same order than deformation of hardened cells under high speed centrifugation, such an agreement seems significant (Notice that the value $\phi_m = 0.59$ has been obtained (BURTON, 1966) for geometrical close packing of discoids modeling underformed RBC). It has been proposed (QUEMADA, 1981) that the value $\phi_M^{NP} \simeq 0.48$ for NP suspension could result from rouleaux packed to form a (loose) tri-dimensional network, the mean mesh length of which containing about ten cells.

(d) Indeed, for suspension the packing concentration $\phi_M^{HA} = 2/k_\infty^{HA}$ — i.e. the packing value one would obtain if cells were packed keeping unchanged their actual state (i.e. keeping their actual effective volume unchanged) — takes the value 0.552 . This value is in close aggreement with the value which has been theoretically found for rigid spheroid suspension (same axis ratio than RBC) and which has been defined as the critical concentration above which "the particle cannot undergo unimpeded rotation because there is unsufficient space to move"(GOLDSMITH and MASON 1967). Furthermore, such a lowering in ϕ_M^{HA} in comparison with $\phi_{Mo}^{NA}=0.61$ can be interpreted as a dilatancy effect, yet observed by SCHMID-SCHONBEIN (1975), on suspension of rigid cells, as shown on Fig. 5.9 . Data fitting of (5.16) for rigid cell suspension exhibits a strong shear thickening effect, $k_\infty^R > k_o^R$, resulting from dilatancy, while normal and crenated cell suspensions show shear thinning effect, nevertheless with $k_o^C < k_o^N$ and $k_\infty^C > k_\infty^N$ i.e. both reduced cell aggregability and deformability of crenated cells compared to normal ones.

(e) Relaxation times $\tau^{NP} \sim 0.2$ and $\tau^{NA} \sim 0.04$ sec are more difficult to be interpreted. As RBC (and a fortiori rouleaux) are too large particles to exhibit any Brownian effects, these times could be thought as Maxwell relaxation times for non rigid particles, $\tau_i \sim \eta_i/G_i$, involving some internal or membrane viscosities η_i and some elastic modulus G_i .

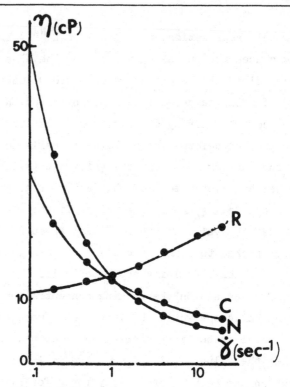

Fig.5.9 Red cells suspended in plasma.
Effect of RBC deformability. Data from
SCHMID-SCHONBEIN (1975). Curves for
fitted values of rheological parameters
as follows.

	ϕ	k_O	k_∞	τ_{sec}	$2/k_O$	$2/k_\infty$
Normal N	0.40	4.88	1.99	1.08	0.41	1.01
Crenated C	0.40	4.56	2.62	0.78	0.44	0.76
Rigid R	0.39	3.23	3.97	1.34	0.62	0.50

Values about 0.2 sec can be obtained from the data of SKALAK (1976) for
shear and dilatation deformations of the RBC membrane. Although single
cell deformability probably dominates the rouleau one, the agreement with
τ^{NP} seems non conclusive, unless the lower value τ^{NA} could be explained
through membrane property modifications associated to albumin molecules
adsorbed in the membrane.

The above discussion shows that close correlations can be expected
between the values of the rheological parameters and the fundamental

determinants of blood rheology, i.e. at given H, essentially RBC aggregation, RBC deformation and plasma content. Some clinical applications will be given in Chap.8.

6. BLOOD AS A THIXOTROPIC-VISCOELASTIC FLUID: EFFECTS OF MECHANICAL AND
   PHYSICO-CHEMICAL PROPERTIES OF THE RBC MEMBRANE.

Up to now, we have been concerned with steady properties of suspensions. Shear-thinning properties that most of concentrated disperse systems exhibit   have been related, in the last section, to dynamic equilibrium between the formation and the destruction of some internal ("microscopic") structure. Moreover, time dependent behaviour observed in the
apparent viscosity not only involves this time dependent "microscopic"
structure (as aggregation or deformation-orientation of particles), but
also,as it will be seen in the next section, the time dependent collective ("macroscopic") structure (as the annular two-phase flow in pipes) /
which results from the development of a cell free layer,close to the cup
(and bob) walls (COKELET et al., 1963), for instance.

Since these processes involve characteristic times (as relaxation
times), such systems will exhibit time dependent properties under unsteady conditions i.e. they are thixotropic materials. However, instantaneous
(reversible) deformation can be supported by the structure, the properties
of which are purely elastic before to be broken. Therefore under steady
conditions, the material possesses both stored energy (by deformation)
and dissipated energy (by creeping flow), what is the general feature
of viscoelastic materials . Additional viscoelastic effects may arise from
elongational deformation and elongational flow. Such effects, which have
been commonly observed in viscoelastic fluid flows through pipes presenting an abrupt reduction in their circular section, can be expected to occur at the branching of two arterioles (compressible flow) or of two
venules (extensional flow), but seem not studied yet.

6.1. Some experimental results about thixotropy and visco-elasticity
of blood and RBC suspensions.

A) Transient experiments (Step function experiments)

As it has been shown in § 1, 2D, superposing thixotropy and viscoelasticity would leave to different flow regimes, depending on the mean
value of shear rate. Such regimes had been observed with blood and RBC

suspensions in coaxial cylinder viscometers. (COKELET, 1972; HEALY and
JOLY, 1975).

Diagramatic representations of these regimes are shown on Fig.6.1

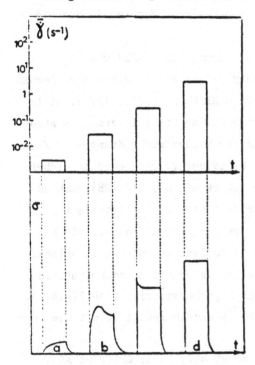

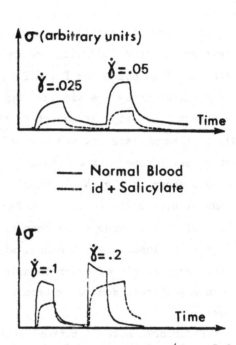

Fig.6.1 Different types of time-
curves of the shear stress in response of applying shear rate steps
with increasing amplitude.

Fig.6.2 Time curves of σ (from J.C.
HEALY and M.JOLY, 1975)

————— normal blood
-------- id. + salicylate

At low values of mean rate ($\dot{\gamma} \lesssim 10^{-2}$-$10^{-1}sec^{-1}$) a relatively slow increase
in σ is observed, that ressembles to what results from retarted elastici-
ty. After stopping the viscometer, stress relaxation occurs (Fig. 6.1a).
Increasing the shear rate ($\dot{\gamma} \lesssim 10^{-1}$-$1sec^{-1}$) leads to a well-known shape
of curve, called an "overshoot", a general behaviour which can be belie-
ved as resulting from superposition of thixotropy and elasticity (i.e.
non-linear visco-elasticity). (See Fig. 6.1b and Fig. 1.5e). At higher
shear rates ($\dot{\gamma} \lesssim 1$-$10sec^{-1}$) thixotropy dominates with, at t = 0, an
abrupt rising of σ up to a peak value, then immediately followed by a

slow decay down to a lower steady value (Fig.6.1c). Finally, at very high shear rates (above 10-50 sec^{-1}), the stress σ rapidly rises to some equilibrium value and remains there until stopping the viscometer, and then quickly drops to zero, what is the common behaviour of pure liquids and solutions.

Comparison of transient behaviour of normal blood (NB) with that of disaggregated blood (DB), with sodium acetylsalicylate to disrupt rouleaux, is shown on Fig. 6.2 a and Fig.6.2 b, (From HEALY and JOLY, 1975). At low $\dot{\gamma}$, the curve shapes are similar, but with an important lowering in steady apparent viscosity. As expected, thixotropy completely disappears from (DB), at intermediate $\dot{\gamma}$, and newtonian behaviour is recovered at higher shear rates. (Comparison of steady viscosities of (NB) and (DB) samples have been shown on Fig. 4.4). Such experiment seem to demonstrate that RBC aggregation would be the main factor responsible of thixotropic behaviour of blood, through the formation of rouleau and of tridimensional network of rouleaux. Nevertheless, since RBC deformation and orientation are also responsible of shear-thinning at high hematocrits (See Fig.4.2), one can expect that some thixotropic effects should result from these processes.

Furthermore, several authors (COKELET et al., 1963; CHIEN et al., 1966: COKELET, 1972) claimed that the interpretation of shear-time curves would require to account for plasma layer formation near the walls of the viscometer gap (and perhaps cell orientation). Such requirement cannot be completely discarded noticely in pathological blood when plasma layer takes shorter times to be formed than in normal blood. Indeed as COKELET (1972) pointed out, many factors are involved in recording the torque-time curve, namely (1) effects due to acceleration of blood from rest, (2) changes in fluid structure as the blood is sheared, i.e. thixotropy and elasticity, (3) dynamic response of the instrument and (4) radial migration of RBC. Moreover, at low shear rates, sedimentation effects can also occur (COPLEY et al., 1976; VINCENT and OLIVER, 1976,). Therefore, a complete interpretation of these time curves, in order to get the relevant flow properties, still remains an open question.

Use of cyclic tests (as a linear increasing of $\dot{\gamma}$ with time, up to a maximum and then decreasing at the same rate down to zero) gives a hysteresis (open) loop, which contains more information than the step changes in $\dot{\gamma}$ (Fig.6.3, BUREAU at al., 1978), and has the advantage to avoid some

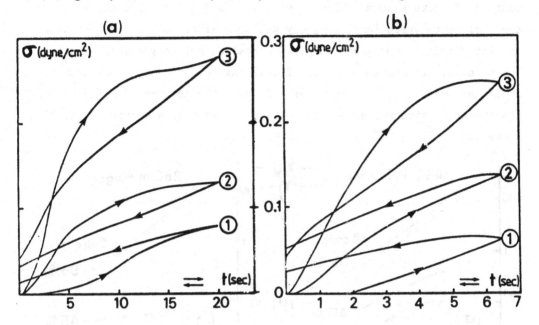

Fig.6.3 Hysteresis loops obtained for linear increase of $\dot{\gamma}$, followed by the same linear decrease, for different rates α
a) t_{max} = 20 sec; $\alpha(sec^{-2})$: 1 = 5.10^{-3} ; 2 = $14.5.10^{-3}$
   3 = 42.10^{-3} .
b) t_{max} = 6.5 sec; $\alpha(sec^{-2})$: 1 = $15.5.10^{-3}$; 2 = 45.10^{-3} ;
   3 = 155.10^{-3} .
(Brom BUREAU et al., 1978).

of the above-mentioned difficulties.

B) Dynamic experimentals (Oscillatory experiments)

Taking a shear rate which varies sinusoidally with time leads to the so-called dynamic response of the system, which can be charaterized by the (complex) dynamical viscosity $\tilde{\eta}$ =η_V - iη_E , η_V and η_E being the viscous and the elastic components, respectively (See § 1, 2 D).

Number of experiments showed that η_V and η_E are functions of the fundamental factors governing blood viscosity — namely hematocrit, RBC

aggregability and deformability, physico-chemical properties of the suspending fluid — and the frequency f of oscillations. For instance, CHIEN
and his colleagues gave the variations of η_V and η_E vs f, at given hematocrit, for RBC in plasma and RBC in Ringer from low frequency measurements in Couette system (CHIEN et al., 1975). Fig.6.4 shows the main features of these variations, noticely the weak effect of cell aggregation
at high H values. On the contrary, at normal H, not only η_V, as expected,
but η_E also, are observed larger if cell aggregation is present than if
it is not present. This likely reflects that the network of rouleaux (in
plasma) has a stronger elasticity than the system of dispersed cells (in
Ringer).

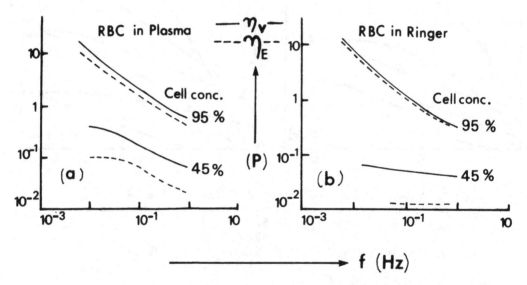

Fig.6.4 Variations of viscous and elastic components of the
complex viscosity, η_V (——) and η_E (———), versus the frequency
f of oscillatory tests, for normal RBC suspended in plasma and
in Ringer (From CHIEN, 1979).

As blood is a shear thinning fluid, attention has been given to
oscillatory measurements in the presence of a given shear rate. Changes
in response reflect similar changes observed in the shear rate dependent
stress-time curves from transient experiments. As an example, Fig.6.5
shows variations of the steady viscosity η_S and the viscous and elastic

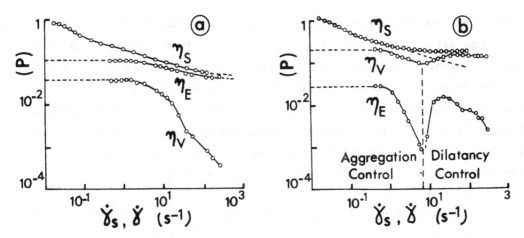

Fig.6.5. Shear rate dependence of steady viscosity η_S, and viscous and elastic components η_V and η_E, of the complex viscosity
f = 2Hz. (a) Normal RBC in plasma, H = 43; (b) Hardened RBC in
plasma, H = 45 (From THURSTON, 1979 a).

components, η_V and η_E , respectively, versus the steady wall shear rate
or the rms value of the oscillatory shear rate at the wall, $\dot{\gamma}_S$ or $\dot{\gamma}$,
respectively (THURSTON, 1979 a). The measurements has been performed at
f = 2 Hz, in normal blood (Fig.6.5a) and suspensions of hardened RBC
(Fig.6.5b), using an oscillatory tube flow apparatus. A spectacular inversion in the elastic component is observed and is interpreted at high
shear rates as resulting from dilatancy effects which originate from cell
hardening. Nevertheless, caution is needed in attempt to understand such
a complex system noticely the sharp increase of end effects that one
would expects as resulting from elongational flows at the entrance and
the exit of the tubes.

Such difficulties in interpretation, and, more generally, in separation of thixotropic and viscoelastic effects, call for improvement of
rheological models for complex materials,nevertheless simple enough to be
available for clinical use, i.e. which avoids unnecessary mathematical
rigour. Some examples of such attempts will be given in the following.

6.2. Some models for thixotropy

As it has been stressed yet, shear-thinning behaviour implies thixotropy, since structure modifications induced by a change in flow condi-

tions need some time, what is measured by a relaxation time.

Many attempts at description of thixotropic behaviour of blood have been proposed. As example, rheogoniometric experimental studies covering a large range of shear rates — from 10^{-3} to 10^{3} sec^{-1} (COPLEY et al. 1973) have been interpreted with the help of a rheological equation proposed by HUANG as a model of isothermal structural change induced by shear stress (HUANG, 1972). For viscometric flow, the shear stress-shear rate relation can be written as a generalized (time dependent) Bingham equation:

$$\sigma = \sigma_0 + [\eta + C\S\beta \; \frac{|\dot{\gamma}|^p}{\dot{\gamma}} \; e^{\int_0^t |\dot{\gamma}|^p dt} \;] \; \dot{\gamma} \qquad (6.1)$$

where σ_0, η, C, $\S$, β and p are characteristic constants as defined in the model. Very fair fittings have been obtained (HUANG et al., 1975).

A phenomenological approach of inelastic reversible thixotropic fluids (CHENG and EVANS , 1965) can be tentatively used. The model is based on two constitutive equations, namely:

(1) a scalar (rheological) state equation:

$$\sigma = \eta(\lambda, \dot{\gamma}) \cdot \dot{\gamma} \qquad (6.2a)$$

is assumed to exist, in which the viscosity depends both on $\dot{\gamma}$ (shear thinning) and on λ , a phenomenological parameter which characterizes the structural state at time t ;

(2) a scalar rate equation:

$$\frac{d\lambda}{dt} = g(\lambda, \dot{\gamma}) \qquad (6.2b)$$

describes the structure changes as a function of both $\dot{\gamma}$ and the structure reached at time t , through the λ value. As an example, the Moore model (MOORE, 1959) is based on the following form of (6.2), with A,B,C as model constants:

$$\sigma = (\eta_1 + C\lambda) \; \dot{\gamma} \qquad\qquad\qquad (6.3a)$$

$$\frac{d\lambda}{dt} = A - (A + B\dot{\gamma}) \lambda \qquad\qquad\qquad (6.3b)$$

Under constant shear rate, the structure reaches an equilibrium state, described by:

$$\lambda_{eq} = \frac{A}{A + B\dot{\gamma}} = \frac{1}{1 + \tau\dot{\gamma}} \qquad (\tau = B/A) \qquad\qquad (6.4a)$$

that gives a shear thinning viscosity (identical to the Willamson's one, See eq.(5.5)):

$$\eta_{eq} = \eta_1 + \frac{C}{1 + \tau\dot{\gamma}} \qquad\qquad\qquad (6.4b)$$

Notice that λ values lie from 0 (fully broken-down structure at $\dot{\gamma} \to \infty$) to 1 (fully built-up structure at $\dot{\gamma} = 0$).

In the case of suspensions of deformable and aggregable particles, a similar approach results from using the intrinsic viscosity k in (5.16) as a structural variable which is governed by the phenomenological rate equation (5.11). Indeed, using the structural parameter:

$$\lambda = \frac{k - k_\infty}{k_o - k_\infty} = \lambda (t) \qquad\qquad\qquad (6.5)$$

transforms (5.11) to

$$\frac{d\lambda}{dt} = A - A (1 + \theta) \lambda \qquad\qquad\qquad (6.6)$$

where $A = \tau_A^{-1}$ and $\theta = \tau_A/\tau_D = (\tau\dot{\gamma})^p$ has been defined in (5.12 - 14). For $p = 1$, (5.6) is identical to (6.3b) from the Moore model. Time behaviour of the system can be deduced from (6.6) and (5.16). For instance if the suspension starts from rest ($\lambda(o) = 1$), the shear stress response to a shear rate step ($0 \to \dot{\gamma}_1$) will be $\sigma = \eta(t) \cdot \dot{\gamma}_1$, where $\eta(t)$ is related, through k(t) in (5.16) or $\lambda(t)$ in (6.5) such as :

$$\lambda(t) = \lambda_\infty + (1-\lambda_\infty) \, e^{-A(1+\theta)t} \tag{6.7}$$

where λ_∞ is the equilibrium value (at $t\to\infty$) of λ

$$\lambda_\infty = \frac{1}{1+\theta} \tag{6.7a}$$

involved in (5.16). Therefore (6.7) gives a correct reprensentation of stress-time curves, as shown on Fig.6.1c. Notice that the relaxation time of the stress is the shear dependent characteristic time for the breakdown of the structure, which appears in the rate equation:

$$\tau_R = A^{-1}(1+\theta)^{-1} \tag{6.8}$$

and which differs from the characteristic time τ involved through θ in the steady viscosity (6.4a). This corresponds to the frequently observed behaviour of the stress which, after cessation of flow under large shear rates, can fall below that stress resulting from small shear rates.

Some difficulties in these models can arise from the too crude approximation introduced by such rate equations, involving only two characteristic times. As for viscoelasticity (see later), description of most of complex fluids requires several characteristic times, sometimes forming a continuous spectrum as in polymer solutions.

More sophisticated models, mainly developped for polymers, have been based on reaction kinetics equations, as rate equations describing the structure formation by linkage of particles (random chain) and the structure destruction by shear induced breaking of links (See, for instance, RUCKENSTEIN and MEWIS, 1973). They could be tentatively used for blood.

6.3 Some models for viscoelasticity

Modelling of linear viscoelasticity is currently performed by use of mechanical models, as more or less complicated combinations of springs and dashpots. The simplest viscoelastic fluid is the Maxwell fluid, represented by a Maxwell element, i.e. a spring (having a Young (elastic)

modulus G) and a dashpot (having a viscosity η) (Fig.6.6.). The corresponding constitutive equation (for stress relaxation experiments) is:

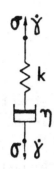

$$\frac{1}{G} \frac{d\sigma}{dt} + \frac{\sigma}{\eta} = \dot{\gamma} \qquad (6.9)$$

In the case of applying a shear rate step from rest, the solution of (6.9) is:

$$\sigma = \eta\dot{\gamma} \, (1 - e^{-t/\tau_M}) \qquad (6.10)$$

MAXWELL element

Fig.6.6. Maxwell
element.

where $\tau_M = \eta/G$ is the Maxwell relaxation
time. Eq. (6.10) defines a stress-time curve in agreement with low shear rate observation when thixotropy has negligible influence.

For a dynamic experiment, applying the oscillating shear rate
$\dot{\gamma} = \dot{\gamma}_1 \cos \omega t$ gives a dynamic response $\tilde{\sigma} = \sigma_1 - i\sigma_2$ that allows to separate the viscous and elastic components of the dynamic viscosity
$\tilde{\eta} = \eta_V - i\eta_E$, respectively:

$$\eta_V = \frac{\eta}{1 + (\omega\tau_M)^2} \qquad , \qquad \eta_E = \frac{\eta\omega\tau_M}{1 + (\omega\tau_M)^2} \qquad (6.11)$$

which involves τ_M through the dimensionless group $\omega\tau_M$.

As the structural viscosity does, a structural elasticity could originate from changes in the kind of structural elements responsible of
elastic properties that would lead to a time-dependent shear modulus G(t).
In fact, as in fluids elastic effects are confined only to the early
part of the time curve, it will be possible in practice to replace G(t)
by its initial value G(o), corresponding to the initial structure of the
system.

Accounting for thixotropy (with non linear viscosity) can be carried
out by adding some elastic effects at short times to the previous shear

thinning model. Characterizing these effects by a Maxwell time τ_M can be estimated by initial values of η and G, $\tau_M = \eta(o)/G(o)$. A model for stress relaxation of a non-linear viscoelastic system then results from generalizing eq.(6.9).

$$\tau_M \frac{d\sigma}{dt} + \sigma = \eta(t)\dot{\gamma} \tag{6.12}$$

where $\eta(t)$ is the time-dependent structural viscosity, as given by (5.16) and (6.6) , and τ_M being estimated from the structure under the initial conditions of the experiment, i.e. $\tau_M = \eta(o)/G(o)$. From (6.12), the shear stress response to a suddenly imposed shear rate step $\dot{\gamma}_1$ will be governed by a competition between the stress building up from retarded elasticity (See (6.10)) and the stress relaxation coming from thixotropy (See (6.6)). Since the relaxation time τ_R, given in (6.8), falls down as $\dot{\gamma}_1$ is increased, the "overshoot" behaviour, mentioned at the beginning of Chap.6 (See Fig.6.1b) will appear as $\dot{\gamma}$ increases. Nevertheless, it must be stress again that complex systems could need a large number of charateristic times.

Accounting for (i) shear thinning and (ii) non linear viscoelasticity, THURSTON (1979b) proposed a very interesting model, based on the Jeffreys'model — combining in parallel N Maxwell elements and a dash-pot (Fig. 6.7.) — which gives a dynamic viscosity (under sinusoidal shear rate):

$$\tilde{\eta} = \eta_\infty + \sum_{\ell=1}^{N} \frac{\eta_\ell}{1 + \omega\tau_\ell} \quad , \quad \tau_\ell = \eta_\ell/G_\ell \tag{6.13}$$

The steady (shear-thinning) viscosity η_S results from (6.13) as $\omega \to o$:

$$\eta_S = \eta_\infty + \sum \eta_\ell$$

To describe thixotropy, as "the manner in which the model elements

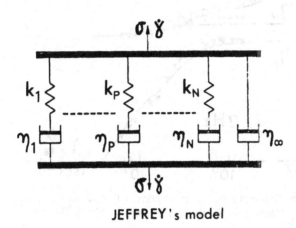

JEFFREY's model

Fig. 6.7.   JEFFREY's Model

change with state of dynamic equilibrium", Thurston assumed that each model element changes with shear rate according to:

$$\eta_\ell = \eta_{o\ell}\, F\, (\dot{\gamma}\, \tau_\ell)$$

$$G_\ell = G_{o\ell}\, F\, (\dot{\gamma}\, \tau_\ell)$$

(6.14)

involving the same "degradation function" $F(\dot{\gamma}\, \tau_\ell)$ which keep each relaxation time τ_ℓ unchanged. In (6.14), $\eta_{o\ell}$ and $G_{o\ell}$ are values at rest.
Further, Thruston proposed to take F in the form!

$$F(\dot{\gamma}\, \tau_\ell) = [1 + (\dot{\gamma}\, \tau_\ell)^2]^{-A}$$

(6.15)

where exponent A will be selected by best fitting of data.
     Fig.6.8 illustrates the results from such a model fitted on data
given in Fig. 6.5a. The model contains 6 relaxing elements — thus 6 relaxations times. A fair agreement – taking into account that data cover
five decades – is observed.

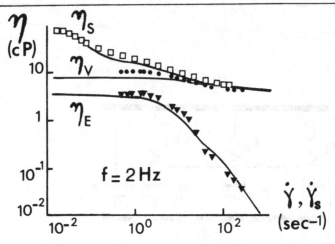

Fig.6.8 Theoretical variations of η_S, η_V
and η_E from the Thurston's model (1979b)

6.4 Effects of mechanical and physico-chemical properties of the RBC membrane.

Properties of RBC membrane, in relation with physico-chemical properties of the suspending fluid (noticely its macromolecular content) are
as evidence involved in thixotropy and viscoelasticity of blood.

Quantitative correlations between these bulk properties and RBC aggregability, on the one hand and RBC deformability, on the other hand,
can be obtained from comparative experiments. We shall limit here to give
some examples:

a) Fig. 6.2a and 6.2b have shown the effect of RBC aggregation upon
thixotropy of blood. Adding sodium salicylate to a blood sample at normal
hematocrit (i) reduces the viscosity at any $\dot{\gamma}$, (ii) suppresses completely the thixotropic behaviour.

b) Fig.6.9, from CHIEN et al. (1975), shows variations of viscoelastic properties in dynamic experiments at f = 0.1 Hz. Differences in viscoelastic behaviour of normal blood (NP) and RBC suspended in albumin
ringer (NA) can be correlated with RBC aggregation, observed by microphotography (COPLEY et al., 1975). As hematocrit is increased, the viscoelastic behaviour of NP tends towards those of NA. They mainly result from
superposition of many factors, noticely mechanical properties of the RBC
membrane and viscosities of intra and extra cellular fluids.

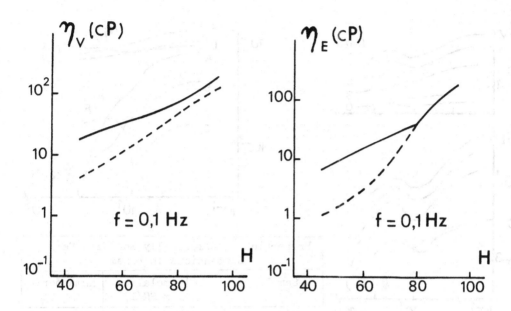

Fig.6.9 Variations of η_V and η_E versus hematocrit for RBC in plasma and RBC in Ringer. (From CHIEN, 1979).(——— plasma, — — — ringer).

c) Fig.6.10, from THURSTON (1979) shows the variations of viscous and elastic components η_V and η_E, versus $\dot{\gamma}$ at f = 2 Hz, during a progressive hardening of RBC over a period of 11 days, leading day after day to a progressive increase of η_V, more especially η_E , which presents a characteristic inversion at about 1 sec^{-1}. Confirmation of this result has been obtained from studying properties of osmotically modified RBC suspended in plasma, where inversion is observed only for crenated cells (i.e. hardened cells) leading to both enhanced η_V and η_E at high shear rates.

Although these unsteady properties of blood are very important since blood circulation is pulsatile, difficulties in their measurement and interpretation show that much work is still needed before satisfactory understanding of the unsteady behaviour of blood will be gained.

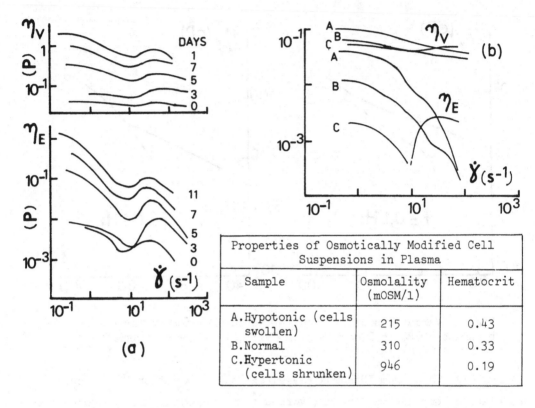

Properties of Osmotically Modified Cell Suspensions in Plasma

Sample	Osmolality (mOSM/l)	Hematocrit
A. Hypotonic (cells swollen)	215	0.43
B. Normal	310	0.33
C. Hypertonic (cells shrunken)	946	0.19

Fig.6.10(a) Variations of η_V and η_E during a progressive hardening of RBC by acetaldehyde. The inversion in η_E can be compared to the one observed in osmotically crenated cells. (b) (From THURSTON, 1979a).

7. BLOOD FLOW IN NARROW VESSELS AS A TWO-PHASE (ANNULAR) FLOW: Blunted
   velocity profile, plasma layer, Fahraeus and Fahraeus-Lindqvist Effects. Two-fluid models.

   As it has been pointed out in §1, it is currently assumed that rheological properties of blood do not influence very much its flow in large
vessels, although some exception may result at low flow rates, near quasi
steady conditions and in the vicinity of changes of the cross section area of the vessel (branching, stenosis or aneurism, ...). On the contrary,
as the diameter of the vessel is lowered, blood rheological properties
appear more and more important from shear thinning, finally complicated
by occurence of phase separation in narrow vessels. However, as in vivo
studies depend on a too large number of unknown variables, one can hope
that precise in vitro analysis of blood flow through narrow tubes will
be able to allow some progress in understanding blood microcirculation
and further, in clinical applications.

7.1. Experimental evidences in blood microcirculation (°).
     Table 7.1 briefly gives main flow characteristics of human microcir-
culation (adapted from SUTERA, 1977).
     Experiments on steady blood flow exhibit four 'anomalous' features,
namely the blunting of velocity profile, the formation of the plasma
layer, the Fahraeus Effect and lastly the Fahraeus-Lindqvist Effect.
     (i) A *blunted velocity profile* is observed near the axis of the
vessel, leading to a plug flow either at high hematocrit or at low value of the vessel to cell diameter ratio, $\xi = 2R/2a$. Such a blunting
is shown on Fig. 7.1 of human nomal blood flowing through a narrow slit
(**DUFAUX** et al., 1980), H = 20 and 40 (the data has been obtained by Laser

----

(°) restricted here to blood circulation in narrow vessels. Nevertheless,
circulation in capillaries shows some analogies through the "axial train"
structure.

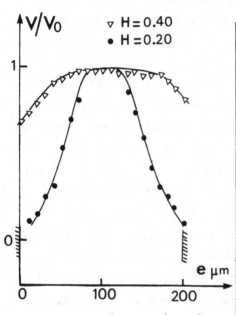

Fig.7.1. Velocity profiles in a narrow slit (1xe=200μmx1.1.cm). Normal blood at H = .20 and .40 (From DUFAUX et al., 1980)

TABLE 7.1. BLOOD FLOW IN HUMAN MICROCIRCULATION

	Vessel Diameter (μm)	Diameter Ratio ε	Blood (°) Velocity (cm/s)
Terminal Arteries	500-2000	60-250	4-20
Small Arteries	70-500	10-60	4
Arteriol.	10-70·	1-10	05
Capilla.	4-10	2-1	01
Venules	10-110	1-14	04
Small Veins	110-500	15-60	2
Terminal Veins	500-5000	60-600	2-10

(°) Average peak values

Doppler velocimetry).

Such results, accounting for the fact that higher flow rates Q values are obtained under the same pressure gradient P, are comparable with similar results on flow of rigid sphere suspensions through tubes (KARNIS et al., 1966). Nevertheless, the independence of velocity profile against flow rate (i.e. linearity of the pressure-flow rate relation) observed by these authors seems to result from a too short range of flow rate variations. Conversely, Fig. 7.2 shows these non-linearity in the P-Q relation, i.e. non-newtonian effects in the apparent viscosity.

(ii) In very narrow tubes ($3 \lesssim \xi < 50$), the existence of a *marginal (plasma) layer* near the wall is currently accepted, leading to consider the flow as a two phase one, with a particle rich axial core surrounded by a particle depleted (or a particleless) wall layer. Such (much debated) existence, is now well-established after Bloch's photographic records

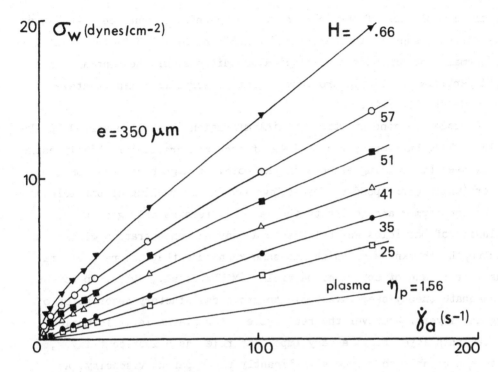

Fig.7.2 Pressure-Flow rate relation of blood flow in a narrow slit
($\ell \times e$ = 1.1 cm × 350 μm) at various hematocrits. σ_w = Pe/2 and
$\dot{\gamma}_a$ = 4Q/ℓe^3 (P = pressure gradient = P/10 cm; Q = flow rate). (From
DUFAUX et al., 1980).

(BLOCH, 1962) of blood flow in various animals. COPLEY and STAPLE (1962)
observed similar features. BUGLIARELLO et al., (1965) carried out measurements in glass capillaries, using anticoagulated normal human blood at different hematocrits, changing the pressure gradient. They found that the
relative layer thickness, δ/R, decreases increasing the hematocrit H and
/or decreasing the wall shear stress σ_w, whatever the radius may be excepted at low hematocrit (H = 28%) for which δ/R appears as very sensitive to
the R value. Values of δ/R are found of about 0.05-0.1 at normal hematocrit.

    Direct studies of single particle motions in Poiseuille flow
(GOLDSMITH and MASON, 1967) showed that rigid particle does not experien-
ce any radial motion if the flow is low   enough, unless the particle
Reynolds number (see eq.(2.4)) is higher than about 10^{-4}, then giving a

radial accumulation of particles near the cylindrical surface at ≈0.5-0.6R
Nevertheless, such a "tubular pinch" (SEGRE and SILBERBERG, 1962) will
not normally occur under physiological conditions. On the contrary, non
rigid particles as liquid droplets migrate towards the axis, whatever is
the value of flow rate.

At moderate concentration, radial migration is still observed (GOLD-
SMITH, 1971), but forces responsible of such a migration are likely coun-
terbalanced by crowding effects in the core. At high concentration the
latter (which correspond to forces involved in the packing of particles)
win on the former and, close to the wall, it remains only geometrical
exclusion of particles which leads to a reduced concentration within an
annulus, the thickness of which has same magnitude that the particle radius in the case of spherical particles (MAUDE et al., 1958). Up to now,
no adequate theory seems available to get a satisfactory description of
marginal layers. Whatever the real process for plasma layer formation
may be, this layer plays a very important role, as a lubricant layer,
which can be able to reduce significantly the apparent viscosity. As
VAND (1949) pointed out, "in the region of high concentrations, considerable slip at the wall might develop due to the layers of low viscosity
along the walls which might finally completely overshadow the effects of
shear inside of the suspension, making the measurements useless".

(iii) A mean (tube averaged) hematocrit H_t is found lesser than the
feed (reservoir) hematocrit H_f. It decreases as the tube diameter diminishes (Fahraeus effect)(FAHRAEUS, 1929).

As COKELET (1976) showed, the discharge hematocrit H_d (i.e. the
average hematocrit of the outflowing blood from the tube) practically
equals the feed hematocrit H_f if $\xi \geq 3$. Below this value, $H_d < H_f$, that
can be thought as resulting from some particle redistribution at the entrance of the tube. For $3 < \xi \lesssim 50$, BARBEE and COKELET (1971) found that
the tube relative hematocrit, $H_r = H_t/H_f$, varies linearly with H_f ,
the slope increasing as R decreases (See Fig.7.3). Similar variations but
with markedly different slopes from the ones expected were obtained with
very narrow tubes (down to 2R = 8.1 μm) (COKELET, 1976). Thus, the follo-

wing approximate relation is obtained

$$H_t \simeq \alpha_1 H_f^2 + \alpha_2 H_f \quad , \text{ with } \quad \alpha_{1,2} = \alpha_{1,2} \, (R) \qquad\qquad (7.1)$$

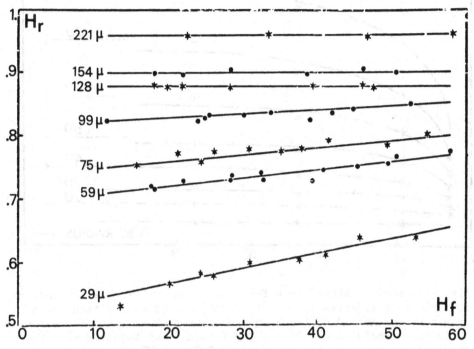

Fig.7.3   Relative hematocrit $H_r = H_t/H_f$ versus H_f for
different tube diameter D. (From BARBEE and
COKELET, 1971).

(iv) the apparent viscosity lowers as the tube diameter decreases (Fahraeus-Lindqvist Effect) (FAHRAEUS and LINDQVIST, 1931).

Number of studies of this effect have been carried out. We shall limit here to display the results of HAYNES and BURTON (1959) in long tubes
(to avoid any entrance effects (L/R $\simeq$ 90) with 67 μm $\leqslant$ R $\leqslant$ 750 μm. They
observed that the higher the hematocrit, the stronger the η_a lowering.
(Fig. 7.4).

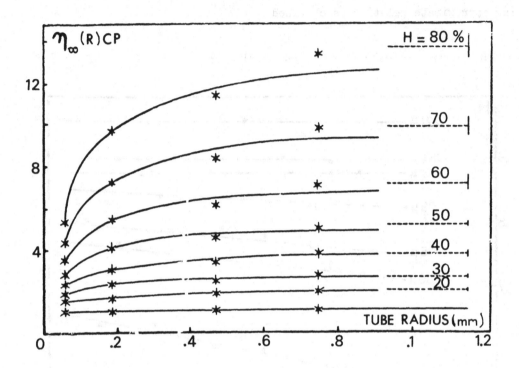

Fig. 7.4 Fahraeus-Lindqvist effect for human erythrocyte suspensions
of various hematocrits at 25.5°C. Smooth curves were fitted to the
points by the method of least squares. Asymptotic values for a tube
of infinite radius are indicated by broken lines together with their
standard errors of estimate. (From HAYNES and BURTON, 1959).

Similar results, expressed as wall shear stress σ_w versus $\overline{U} =$
$Q/\pi R^3 = \dot{\gamma}_a/4$ (where $\dot{\gamma}_a$ is the apparent shear rate (see eq. (2.12))ha-
ve been obtained by BARBEE and COKELET (1971). Fig. 7.5 shows the data
obtained (i) with a large tube (where no phase separation occurs, hence
$H_t = H_f$.(ii) with a narrow tube, 29 μm in diameter, where $H_t \neq H_f$ varies
as a function of H_f, as indicated. Comparing large and small tubes data
at $H_f = 0.559$, for instance, shows Fahraeus-Lindqvist effect as the lowering in σ_w at any $\overline{U}$ value. These authors found an important empirical
fact from their data : the relation σ_w versus $\overline{U}$ for large tubes fits the
small tube data at a given feed hematocrit H_f, provided the tube hemato-
crit H_t was used in the relation σ_w $(\overline{U}, H)$ instead of H_f as in large tube

that is :

$$\sigma_w\ (\overline{U}\ ,\ H_f,\ R)\ \simeq\ \sigma_w(\overline{U}\ ,\ H_t\ ,\ \infty) \tag{7.2}$$

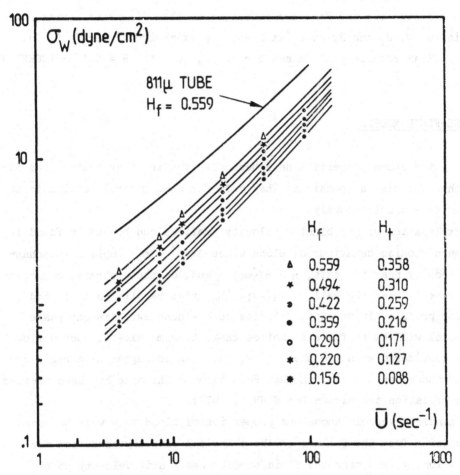

Figure 7.5 Relationship between wall shear stress and average blood velocity divided by the tube diameter. The points are experimental data obtained with 29μm tubes : the curves represent data obtained with an 811 μm tube. (From BARBEE and COKELET (1971))

where $H_t = H_t(H_f,R)$ is given in (7.1). For example (Fig. 7.5) the curve σ_w vs $\overline{U}$ for large tube (about 800 μm) at hematocrit $H = 0.358$ fits remarkably well the small tube data with feed hematocrit $H_f = 0.559$. Therefore one can assert that, empirically, the main part of the Fahraeus-Lindqvist effect results from the Fahraeus effect, i.e. the wall shear stress, as function of $\overline{U}$, H_f and R, must (at least approximatively) verify (7.2). Again similar results hold in smaller tubes, down to 2R = 8.1 μm (COKELET 1976).

7.2 ONE-FLUID MODELS

All the above properties have been observed in other systems, as rigid sphere or disc suspensions. They call for some general model able to explain them simultaneously.

Explanation of the blunted velocity profiles can be easily found in the non-newtonian behaviour of blood since $\partial\eta/\partial\dot{\gamma} < 0$ implies an enhanced viscosity near the axis, $\eta \simeq \eta(\dot{\gamma}=0)=\eta_0$ and, on the contrary, a decreased viscosity near the wall, $\eta\simeq\eta(\dot{\gamma}=\infty) << \eta_0$ that results to a blunted velocity profile. In terminal arteries such a non-newtonian one phase flow model will be sufficient. Notice that, to some extent, such a flow can be considered as a two phase flow, the plugged region as a rigid core, surrounded by a fluid annulus. Formation of the plug has been treated as a percolation transition (de GENNES, 1979).

In narrow vessels anomalous properties of blood flow were believed as resulting from the failure of the classical boundary condition for fluid velocity, assuming the fluid studied had a slip velocity at the wall (See for ex. JONES, 1966). Another explanation, as the non validity of continuous description, is found in the so-called sigma-effect, where the flow through the tube is considered as the summation of contributions from a finite number of coaxial layers of finite thickness (DIX and SCOTT BLAIR, 1940).

7.3 - TWO-FLUID MODELS

The above models failed to give an overall and self-consistent explanation of velocity profiles as well as the remaining experimental features and nowadays two-fluid models appear as the more credible. The decrease of concentration near the wall exagerates the blunting of $v(r)$ since, in addition to $\partial\eta/\partial\dot\gamma < 0$, one has $\partial\eta/\partial H > 0$. Thus a model of blunted velocity profile will be very easily found.

Although the following results hold in other geometries (as narrow slits, for instance, QUEMADA et al, 1980) we will limit our discussion to flow through narrow tubes, taking again as a two-phase model, a core ($0 \leqslant r \leqslant \beta R$, subscript s) surrounded by a peripheral layer ($\beta R \leqslant r \leqslant R$, subscript w). In order to modelize the particleless (plasma) layer, we shall take $H_w = 0$. Some "classical models will be discussed in the following subsections.

7.3.1 - The simplest model is formed by two newtonian fluids, with constant viscosities $\eta_s = \eta(H_s)$ and $\eta_w = \eta_p$

With appropriate boundary conditions (no slip velocity at the wall, $v_w(R) = 0$; continuity of shear stress at $r = \beta R$, $\sigma_w(\beta R) = \sigma_s(\beta R)$; finite velocity gradient on the axis, $(\partial v_s/\partial r)_o = 0$), solving Navier-Stokes equations for the two-phases, allows to calculate the total volume flow rate (BAYLISS, 1952 ; HAYNES, 1960 ; THOMAS, 1962) as the sum of core and wall layer ones

$$Q = Q_s + Q_w \tag{7.3}$$

with

$$Q_s = A \left[2 \beta^2 (1 - \beta^2) + \beta^4 \frac{\eta_p}{\eta_s} \right] \tag{7.3a}$$

$$Q_w = A(1 - \beta^2)^2 \tag{7.3b}$$

where $A = \pi R^3 \sigma_R / 4\eta_p$ and $\sigma_R = \sigma_w$ (R) is the shear stress at the
wall. The apparent viscosity, η_a defined from the Poiseuille's law,
$Q = \pi R^3 \sigma_R / 4\eta_a$ is then given by :

$$\eta_a = \eta_p \left[1 - \beta^4 (1 - \frac{\eta_p}{\eta_s}) \right]^{-1} \tag{7.4}$$

Assuming the plasma layer thickness δ much smaller than R (what is
observed at normal hematocrits), i.e. $\delta/R = 1 - \beta \ll 1$,   , leads to
the following good approximation of (7.4.) :

$$\eta_a \approx \eta_s \left[1 - 4 \frac{\delta}{R} (\frac{\eta_s}{\eta_p} - 1) \right] \tag{7.5}$$

Since δ mainly depends on particle-wall interaction, its value, in
the limit $\delta \ll R$ should be very close to the one obtained neglecting
wall curvature, thus leading to assume that δ does not depend on R. As
a consequence (7.5) gives the apparent viscosity as a decreasing function
of (1/R). Therefore (7.5) describes the Fahraeus-Lindqvist effect, as
$(\partial \eta_a / \partial R) < 0$.

Fig. 7.6 , from MIDDLEMAN (1972) displays fitting of eq.(7.5) on
the data of Fahraeus-Lindqvist (1931). In tubes with relatively large
diameters, a linear variation is obtained, and η_s/η_p and δ can be deduced from (7.5) as $\eta_a = f(1/R)$. Values of δ about 0.7 µm are found from
this data, and appear as limiting values at high wall shear stresses
(σ_w > 80 dynes/cm^2) and at normal hematocrit (which were not measured).

Fitting eq. (7.4) on the data shown in Fig. 7.5., HAYNES (1960) obtained the curves in Fig. 7.5. He calculated δ for each H value : in
the normal range of H, he found $\delta \approx 3$ µm. As evidence, such values, indirectly calculated through a model, are very sensitive to the choice of

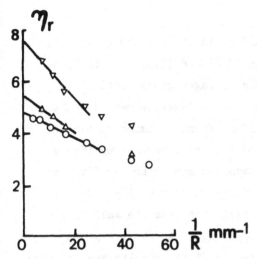

Fig.7.6 Data of Fahraeus-Lindqvist (1931) plotted as a test
of eq.(7.5). (From MIDDLEMAN,
1972).

the model and to the precision of
pressure flow rate measurements.

7.3.2 Discarding the assumption
that the core behaves as a newtonian
fluid, THOMAS (1962) gave a self-
consistent treatment of the problem.
He obtained formally the same equations than (7.3), (7.4), but where
η_s is the apparent viscosity of the
core, defined as if it was flowing
through a fictive tube of radius
βR, with a slip velocity $v_s(\beta R)$
and a shear stress $\sigma_s(\beta R) = \beta\sigma_R$
at the fictive wall. Written with
the help of relative fluidities, $F = \dfrac{\eta_p}{\eta}$, eq. (7.4) becomes :

$$F_a = 1 - \beta^4 (1 - F_s) \qquad (7.6)$$

where F_a = apparent relative fluidity for feed hematocrit H_f and under
wall shear stress σ_R and F_s = core relative fluidity, which depends on
core hematocrit H_s and shear stress $\beta\sigma_R$. Moreover, Thomas accounted
for Fahraeus Effect which had been neglected in previous works.
From equation of continuity of the suspended phase :

$$H_f\, Q = H_s\, Q_s = H_f (Q_w + Q_s) \qquad (7.7)$$

it follows, with (7.3a) :

$$\frac{H_s}{H_f} = \frac{Q}{Q_s} = \frac{1 - \beta^4 (1-F_s)}{2\beta^2(1-\beta^2)+\beta^4 F_s} \qquad (7.8)$$

If the function $F = F(H, \sigma)$ for fluidity in the absence of wall effects is available, the only unknowns in (7.6) and (7.8) will be H_s and β. Thomas (uselessly) introduced the tube averaged concentration $H_t = \beta^2 H_s$ and solved eq. (7.6) and (7.8) by a relaxation method, taking successive approximations to the value of β. Then, assuming σ_R high enough to recover newtonian properties for the core, and taking (See eq. (3.14) $F = 1 - kH$ for core fluidity, he applied his method to different data. For rigid spheres of radius a, he found $(\delta/a) = 0.71 - 0.76$, in satisfactory agreement with geometrical exclusion near the wall. For suspensions of mammalian red cells, δ was found equal to about 4 μm, however decreasing or increasing as R is lowered. These results are not completely convincing ones, in one part because of the incompleteness of the data (hence the approximation that Thomas made) and in other part because of the lack of non-newtonian effects in the core ($\dot{\gamma}$ going to zero on the axis).

7.3.4. A non-newtonian behaviour for the core has been taken into account by CHARM and KURLAND (1962) and by CHARM et al. (1968), using a power law fluid and later a Casson fluid eq.(5.1). They calculated the pressure-flow rate relation, that reduces in our notations to the apparent relative fluidity such as :

$$F_a = 1 - \beta^4 (1 - F_s) + G(\beta, \frac{\sigma_0}{\sigma_R}) \qquad (7.9)$$

which differs from (7.6) by an additional term G which vanishes as yield stress σ_0 tends to zero ($^\circ$). From the existence of σ_0, a partially plug flow takes place, the radius of which is $R_{Plug} = R\sigma_0/\sigma_R$. Eq.(7.9) was

($^\circ$) In the limit of $x = \sigma_0/\sigma_R \ll 1$, one has $G(\beta, x) \simeq \frac{4}{3} \beta^3 x [1 - \frac{6}{7}(\frac{\beta}{x})^{\frac{1}{2}}]$

used by CHARM et al., to analyse their pressure-flow rate measurements
in a large number of capillaries of different sizes. Independently, the
Casson parameters K and σ_0, entering in Eq. (5.1) were determined as
functions of H by Couette viscometry. Furthermore, radial distribution
of cells, H(r) was assumed to be similar to the one observed by PALMER
(1965) in rectangular tubes. Fitting (7.9) on a data relative to flow
under a wide-range of pressure gradients and flow rates, through a large
number of tubes, different in size, led CHARM et al to find results in
good agreement with those from in vivo measurements (See table 7.2).
They deduced variations of the ratio $\delta/2R$ vs H which appears scattered,
that would be due to (i) the failure of Casson law at low shear rates
and/or (ii) the choice of Palmer's radial distribution H(r), which is
questionable since very abrupt profiles have been deduced from analyzing
viscometric data (WATANABE et al., 1963).

TABLE 7.2 OBSERVED MARGINAL PLASMA LAYERS IN FROG MESENTARY VESSELS
AND CALCULATED FROM FLOW OF HUMAN BLOOD IN GLASS TUBES.

(BLOCH, 1962)		(CHARM et al., 1968)	
From mesentary (High speed photography,ϕ unknown		Human blood in glass tubes (Calculated from pressure-flow rate) $\phi = 0.3-0.45$)	
Inside diameter (μm)	δ/D	Inside diameter (μm)	δ/D
71-80	0.0416	71.8	0.0440
141-160	0.0280	155.0	0.0220
241-270	0.0230	256.0	0.0214

7.3.5. Alternatively, the non-newtonian viscosity equation (5.16) has
been used to describe the core behaviour, keeping (7.8) unchanged. Rheological parameters entering in (5.16) were determined with the help of a
low shear coaxial viscometer. The core hematocrit was calculated, after

elimination of F_S between (7.6) and (7.8) :

$$\frac{H_s}{H_f} = (1 - \frac{(1 - \beta^2)^2}{F_a})^{-1}$$

(7.10)

For given R, σ_R and H_f, experimental value of fluidity, F'_a can be obtained. Putting $F_a = F'_a$ into (7.10) leads to :

$$H_S = H_S(\beta)$$

(7.11)

Successive approximations for the value of β (as in the THOMAS' method) are taken in (7.11), allowing to calculate F_S (H_S, $\beta\sigma_R$) then putting into (7.6), until the difference between the calculated value and the experimental one, $F_a - F'_a$, becomes unsignificantly different from zero. Finally, after the value of β have been obtained for a given σ_R the corresponding theoretical velocity profile can be calculated and compared to the experimental one, measured through Laser-Doppler-velocimetry. (DUFAUX et al., 1980, See Fig. 7.1).Such a model has been applied to narrow slits (thickness = h). Using again σ_R for the wall shear stress $\sigma_R = \sigma_w(h)$,Fig.7.7 gives variations of F'_a and of $\delta = h(1 - \beta)$, as functions of σ_R (QUEMADA et al., 1980). It shows that, as σ_R increases from zero, δ increases, reaches a maximum and then decreases and seems to tend towards zero at very high σ_R (limiting value F_∞ was used as $\sigma_R \rightarrow \infty$). Similar features has been found at different feed hematocrit and slit thickness. Although these observations were made in narrow slits, it can be expected that no dramatic differences would exist between slits and pipes. Indeed, such a maximum on the curve $\delta = \delta(\sigma_R)$ has been observed in 200 μm diameter tube, by microphotography under dark field illumination (DEVENDRAN and SCHMID-SCHONBEIN, 1975).

Whatever the model may be, the importance of the (lubrificating) plasma layer requires further work before significant conclusion about its physiological importance would be drawn.

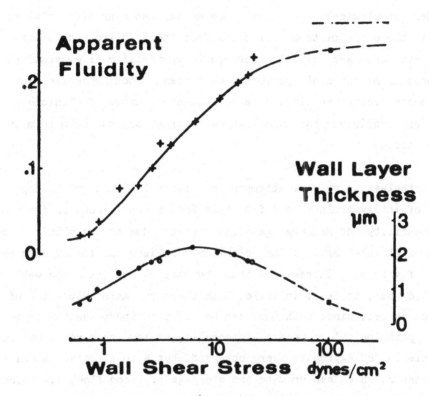

Fig.7.7(a)Apparent fluidity F_a' and (b) Wall layer thickness δ vs.
wall shear stress σ_R . Normal Human Blood Flow (ϕ_a = .57) through a
slit (350 μm x 1.1cm x 10cm).T=23°C. Rheological variables of the
same blood sample (from Couette Viscometry) :k_0=3.24, k_∞=1.68,
τ=.135sec.(Points O: from fitted curve on F_a' data; Points □:from
theory).

8 – <u>CLINICAL HEMORHEOLOGY</u> – <u>CONCLUSION</u>

Applying the analysis developped in the previous sections to blood
flow under pathological conditions comes up against many difficulties,
mainly (i) the extrapolation of results from the in vitro conditions
to the in vivo ones and (ii) the increase in variability (i.e. standard
deviations) of measurements performed on patient blood, sometimes leading to contradictory resalts. As a consequence, number of clinicians
are not very convinced that *quantitative* hemorheology can help them in
their practice.

Nevertheless, number of attempts at interpretation of microscopic
features of intravascular blood flow have been recently done in terms of
RBC deformability and RBC aggregability. Indeed, due to the clinical importance of platelet aggregation, clinicians believe the former more responsible of viscosity increasings than the latter. Now, it seems well
established that, at least in vitro, both these two main properties of
RBC are concerned, since both they can be affected independently or together by pathological processes. Especially, at low shear rate, not only
high levels in RBC Aggregation are observed, but also elevated values in
yield stress σ_0 which can promote the stoppage of blood flow, i.e. blood
stasis. Many examples of application exist but only some of them will be
recalled hereafter.

<u>Steady viscosity measurements.</u>

Poor blood flow, instead of being only considered as resulting from
heart failure or from a blood vessel disorder, now appears as resulting
from high values of blood viscosity, which is observed in many diseases.
However, several attempts to correlate enhanced whole blood viscosity
and/or plasma viscosity to different variables,as H and η_p ,in various diseases did not give very convincing results (See for ex., ISOGAI et al.,

MATSUDA et MURAKAMI, 1976 ; DINTENFASS and KAMMER, 1977), although si-
gnifiant increases in viscosity have been found, especially at low shear
rates (For ex., see DINTENFASS, 1977, 1979). DORMANDY et al. (1973a)
studying patients with intermittent claudication, showed that blood
viscosity of these patients was significantly raised, compared to normals.

DINTENFASS (1971) gave an intensive study of the high viscosity syn-
droma in various diseases (ischaemic, sickle cell, hemolitic anemia,
polycythemia, shock .....) (See Table 8.1 , after DINTENFASS et al,
1966 b).

Fibrinogen concentration plays the most important role in RBC Aggregability. Its lowering (afibrinogenaemia) leads to the recovery of newtonian properties and the lack of yield stress (MERRILL, 1969). Its increasing leads to enhanced viscosity, as in heart failure (KELLOG and
GOODMAN, 1960). Similar effects can be expected to result from the presence of abnormal proteins, as macroglobulins (Waldenstrom's disease) or
IgG1 myeloma (LINDSLEY et al., 1973). Polycythaemia remains a more obs-
cure disorder which is probably resulting from complex association of
the increase in Hematocrit and several factors. Same difficulties occur
in heart attack, in shocks ... Abnormal increase in viscosity of diabe-
tics have been shown, especially if they are associated with abnormal
circulation in retina (HOARE et al., 1976). Fig. 8.1

Changes in RBC Deformability, as in sickle cell disease lead again
to increase in viscosity at low oxygen pressure (i.e. de-oxygenated RBC).
(See CHIEN et al ; 1970). Changes in RBC shape (from pH modifications
for instance), in spherocytosis or in parasitosis (as malaria) also give
an enhanced viscosity.

Table 8.1

BLOOD VISCOSITIES IN NORMAL CONTROLS AND IN PATIENTS SUFFERING FROM
THROMBOTIC AND OCCLUSIVE DISEASES

Shear rate sec^{-1}	Viscosity of blood in poises		
	Normal women	Normal men	Patients
0.01			
$\bar{X}$	11.9	9.9	57.5
$\bar{X} \pm SD$	4.9–28.6	4.4–22.1	21.7–157
$\bar{X} \pm 2SD$	2.0–75	1.9–49.4	7.8–428
0.1			
$\bar{X}$	2.22	2.10	7.23
$\bar{X} \pm SD$	1.23–3.94	1.13–3.93	3.15–16.70
$\bar{X} \pm 2SD$	0.69–7.00	0.60–7.35	1.36–38.3
1.0			
$\bar{X}$	0.41	0.45	1.00
$\bar{X} \pm SD$	0.28–0.59	0.27–0.82	0.60–2.00
$\bar{X} \pm 2SD$	0.19–0.87	0.15–1.41	0.24–4.00
7.2			
$\bar{X}$	0.13	0.15	0.22
$\bar{X} \pm SD$	0.09–0.18	0.11–0.22	0.16–0.32
$\bar{X} \pm 2SD$	0.06–0.26	0.08–0.31	0.11–0.45
29			
$\bar{X}$	0.07	0.095	0.106
$\bar{X} \pm SD$	0.06–0.08	0.07–0.13	0.08–0.14
$\bar{X} \pm 2SD$	0.05–0.09	0.06–0.17	0.06–0.19
118			
$\bar{X}$	0.048	0.056	0.065
$\bar{X} \pm SD$	0.045–0.051	0.048–0.065	0.041–0.101
$\bar{X} \pm 2SD$	0.043–0.054	0.041–0.076	0.026–0.158

(From DINTENFASS et al., 1966 b).

Number of the above-mentioned studies remained only qualitative because they used averaged values of viscosity from large series of normals and patients. Although systematic increase in mean viscosities are measured one observes some overlapping of normal and patients values (See Table 8.1). These findings result from superposition of viscosity changes with variations of rheological variables as H, n_p, RBC Aggregation and RBC Deformation ... for each **donor** , on the one hand, and on the other hand, to viscosity variations when one passes from one patient to another. In fact, getting on a quantitative study needs to reduce data from each **donor** before to make some averaging procedure. CERNY et al. (1974) gave blood viscosity of patients with sickle cell trait disease. Fig.8.3 shows viscosity variations for normal blood (HbAA) for two age groups of people (young, 12-29 years old ,and old ,69-91 years old) as well as for a sickle cell anemia (HbSS) (from CHIEN et al., 1970). Fig. 8.2 also contains the viscosity-shear rate curves for sickle cell trait blood (HbSA) for five young people (4-41 years old) and an elder person (61 years old), which shows an increase by at least 49 %.

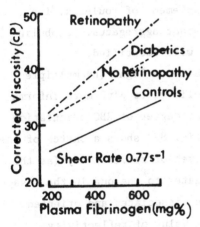

Fig. 8.1 Relationship between viscosity and plasma fibrinogen (from HOARE et al, 1970)

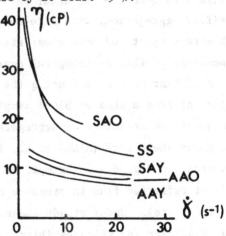

Fig. 8.2 Comparison of viscosities of different bloods.
SS - Sickle cell anemia
SAO - Sickle cell trait-old
SAY - Sickle cell trait-young
AAO - Normal old ,AAY Normal young.
(from CERNY et al., 1974).

Moreover, at low shear rates, the latter exhibits a variation very close
to the SS one.

SCHMID-SCHONBEIN and WELLS (1971) claimed that viscosity cannot be
sufficient for relevant clinical studies since "even the most sophisticated rotational viscometers, while capable of measuring bulk viscosity
near stasis, are subject to artefacts introduced by the material under
study and present the net effect of several different mechanisms". Therefore, one needs to employ simultaneously different tools and methods a-
mong them these authors proposed the "Rheoscope", a transparent counter-
rotating chamber, which can be used to quantify the kinetics of RBC aggregation from light transmission analysis.

The rheoscope has been used in an extensive study of pathological
RBC aggregation (blood sludge) (SCHMID-SCHONEBEIN et al., 1973). Such a
method allowed photometric quantification of the kinetics of the RBC
Aggregation, in relation with changes in viscosity, according to values
given in Table 8.2.
While normal RBCs physiologically aggregate to form rouleaux, with end-
to-site attachment, RBCs, under pathological conditions, exhibit an intensified aggregation, with side-to-side attachement of rouleaux, i.e.
both more compact and more shear stress resistant aggregates. Globulin
responsible of this pathological aggregation were identified.

A similar method, but using the back-scattered light (by multiple
reflexion) from a sheared blood sample, is believed to give more information about the structure of aggregates and the degree of RBC orientation
at a given shear rate (MILLS et al, 1980). Fig. 8.3 shows a sketch of the
apparatus and Fig. 8.4 an example of time curves of reflectivity (as the
ratio of reflected flux in presence of aggregates to the one in the disaggregated state). Under steady shear, RBC are dissociated and oriented
(Fig. 8.4, part (a) which exhibits a constant value of reflectivity).
After abrupt stopping of the cup viscometer, random RBC orientation is
quickly reached (peak b) and then, RBC stack to form rouleaux at $\dot\gamma = 0$,
‹leading

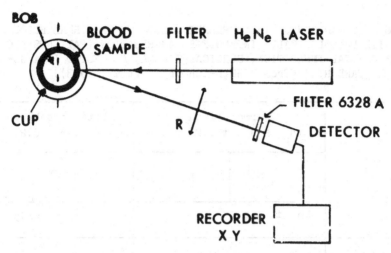

Fig. 8.3.   Sketch of the apparatus for measurement of intensity of the
             light back-scattered by the suspension (from MILLS et al,
             1980).

to aggregation times about 5 sec. for normal blood and much higher values
for pathological blood.

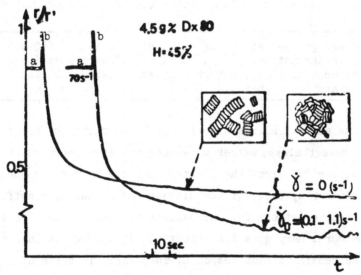

Fig. 8.4.   Variations of the reflectivity of the suspension versus time.
             (N) Normal blood.   (P) Pathological blood. (from MILLS et al,
             1980).

TABLE 8.2 COMPARISON OF THE APPARENT VISCOSITY OF NORMAL BLOOD AND OF
BLOOD WITH PATHOLOGICALLY INTENSIFIED AGGREGATION (PATIENTS RECOVERING
FROM MYOCARDIAL INFARCTION, SURGICAL TRAUMA) AS TESTED IN A GDM-VISCOME-
TER WITH GUARD RING (from SCHMID-SCHONBEIN et al, 1973).

	Normal blood $(\eta_p = 1.2 \pm 0.05$ cP$)$		Patient's blood $(\eta_p = 1.46 \pm 0.20$ cP$)$	
	η_a(cP)	η_r	η_a(cP)	η_r
Hematocrit	40 $\pm$ 1 %	n = 22	n = 15	
0.1 sec^{-1}	42.6 ± 11.4	35.5 ± 9.5	56.18 ± 8.8	38.8 ± 6.2
1.0 sec^{-1}	13.8 ± 3.3	11.5 ± 2.7	17.8 ± 3.6	12.1 ± 1.9
20.0 sec^{-1}	5.0 ± 0.66	4.15 ± 0.55	5.2 ± 0.5	3.59 ± 0.5
Hematocrit	45 $\pm$ 1 %	n = 20	n = 12	
0.1 sec^{-1}	67.7 ± 12.5	56.5 ± 10.4	71.76 ± 13.22	49.02 ± 6.41
1.0 sec^{-1}	17.3 ± 3.2	14.4 ± 2.6	23.10 ± 4.69	15.39 ± 2.13
20.0 sec^{-1}	5.8 ± 0.74	4.8 ± 0.062	6.42 ± 0.88	4.3 ± 0.43

The values are grouped according to hematocrit values (40 ± 1 %,
45 ± 1 %). In the patient group, low shear viscosity is indeed higher
than in the control group, but so is plasma viscosity. Relative apparent
viscosities are not higher.

All the above findings underline the importance of viscosity in
diseases. Nevertheless, *values* of viscosity, even measured at a standard
hematocrit, for different shear rates -especially for low ones - are very often neither specific of the disease under study nor sufficiently
free of large variations for one patient to another one with same disease. Furthermore, many questions about blood circulation in arterioles
and venules remain without response. Therefore, in spite of the fact that
blood complexity asks for the possibility to construct a rheological
model for blood, such a model appears very necessary, even if it is an
approximate one. As too much parameters enter in the problem, one must
try to adopt a more general viewpoint, considering as in the previous

sections that blood behaves as a high concentrated suspension which exhibits shear-thinning from RBC Aggregation (RBCA) and Deformation (RBCD).
Applying general rheological laws - as Casson law for instance - is limited since these two fundamental processes (RBCA and RBCD) are not simply
contained into specific parameters. On the contrary, through the intrinsic viscosity given by eq. (5.16) assuming again all changes due to pathology are included in it (°), these processes can be taken separately
into account by the single parameters k_0 and k_∞. When compared to normal
values, k_0^n and k_∞^n , respectively, variations of k_0 and k_∞ can be expected
to reflect variations of RBCA and RBCD, respectively. Tables 8.3 and 8.4

Table 8.3 Determination of Rheological Parameters from

Δ   data of SCHMID-SCHONBEIN et al (1973)

0   data of SCHMID-SCHONBEIN et al (1971)

+   data of USAMI et al (1971)

	ϕ	η_f cP	k_∞	k_0	sec^{-1}
Normal blood Δ	0.40 0.45	1.2 1.2	1.84 2.07	4.65 4.33	2.23 1.88
Sickle cell anemia 0	40	1.4	2.83	4.63	4.94
Packed cells +	0.88	1.2	1.83	2.26	2.16

(°) Such an assumption means that the form 3.17 for the relative viscosity holds again for pathological blood.

depict such variations, using various literature data. Values of Rheological Parameters (RP) are given on Table 8.3 for :

(i) Normal blood, at ϕ = 0.40 and 0.45 (from data of Schmid-Schonbein et al. 1973).

(ii) Sickle cells, i.e. RBC having abnormal Hemoglobin, hardened when de-oxygenated, but which recovers some flexibility when oxygenated. (from data of SCHMID-SCHONBEIN and WELLS 1971).

(iii) Packed cells, at ϕ = 0.88 (from data of USAMI et al. 1971).

On Table 8.4, comparison between η (or $1/\eta$) calculated from these RP-values, and those directly observed by SCHMID-SCHONBEIN et al. (1971) is shown.

Table 8.4 Comparison of observed and calculated apparent viscosities
- observed values from measurements of SCHMID-SCHONBEIN et al (1971).
- calculated values from eq. (5.16) (with r = 1/2) from RP values given in table 8.3.

	Obs.	calc.	obs.	calc.
	\multicolumn{4}{}{apparent viscosity (cP)}			
Normal Blood Δ	at $\dot{\gamma}$ = 0.1 sec 42.6 ±11.6 67.3 ±12.3	35.3 68.2	at $\dot{\gamma}$ = 230 sec 3.6 ± 0.3 4.1 ± 0.5	3.0 4.1
Sickle Cell anemia 0	at $\dot{\gamma}$ = 0.1 sec 68 ± 19	70.8	at $\dot{\gamma}$ = 230 sec 5.1 ± 0.6	6.64
	\multicolumn{4}{}{apparent fluidity (Pois?^{-1})}			
Packed cells +	at $\dot{\gamma}$ = 2.30 sec 0.61	0.86	at $\dot{\gamma}$ = 230 sec 1.51 (observed at ϕ = 0.95*)	2.64

* In the third case, RP-values are not known at ϕ = 0.95 used in observations. Then, calculated values of fluidity, for ϕ = 0.88 are somewhat higher than the observed ones.

As expected, RP-values appear as concentration dependent. Nevertheless their determination, at fixed ϕ ,   has been proposed as a general rheological characterization (QUEMADA, 1976 b).

Large increases in RBCA have been found by fitting the data of SCHOLZ
et al (1975) for two groups of critically ill patients : Group I = patients suffered from violent trauma or operative injury; Group II = patients with generalized septis.Table 8.5 shows significant increases in
k_o (from enhanced RBCA) and slight increases in k_∞ (which can be related
to RBCD lowering as a consequence of effects of transfusion of stored
blood to several patients). Dramatic increase in values of yield stress
$\sigma_0 = \eta_\infty \tau$, close to the values of the Casson's one (See eq.(5.1)),i.e close
to packing in the core, have been interpreted as increase in cohesion
of RBC Aggregates that could traduce prethrombotic situation for these
patients (QUEMADA, 1976a).

TABLE 8.5 : Enhanced RBC Aggregation

	H	η_p (cP)	k_o	k_∞	τ sec	$\sigma_0 = \eta_\infty / \tau$ (dynes/cm^2)	H_c
GrI	32.2	1.28	6.24	2.08	1.50	0.0193	32.1
GrII	30.2	1.36	6.49	2.04	1.54	0.0185	30.8
Normal	42.6	1.19	4.33	1.61	7.80	0.0035	46.2

Moreover k_o -values found for patients are critical values, in the
sense that each corresponding critical (packing) value of hematocrit,
H_c = $2/k_o$ is very close to the actual hematocrit of patient, that again
represents a very dangerous clinical situation, since blood stasis can
then promote complete stopping of blood flow (i.e. a shock) (QUEMADA,
1976 a).

Once a viscosity equation is available, models of blood circulation
in narrow vessels, taking into account the presence of the plasma layer,

can be constructed, if one knows independently the Fahraeus Effect, i.e. the changes of hematocrit in the vessel, compared with the feed hematocrit. Such models should be able to explain the role that RBCA plays in the microcirculation, as L.E. Gelin early showed in a very illuminating experiment using a X branching tube (GELIN, 1961).

Further aspects, especially physiological ones, have been given in a recent review by 3CHMID-SCHONBEIN et al. (1979).

Unsteady measurements

Number of transient or dynamic measurements have been performed on pathological blood. (See for instance LESSNER et al., 1971 ; BUREAU et al., 1976; COULTER and SING, 1974 ; THURSTON, 1976 ; CHIEN et al., 1975).

Clinical interest of such studies are evident but difficulties to get available models render still incomplete the understanding of experimental data in terms of fundamental processes.

Some concluding remarks

In these lectures, I tried to analyse in detail some models as tools to reach a comprehensive level in analysis of viscometric data. Indeed, data analysis can be only performed with the help of models, which must be both sufficiently complicated for significant describing of complex fluids as blood, and simple enough to be of practical use, until clinical routine. One has to hope improvements in the clinicians'belief in the usefullness of models.

Macroscopic behaviour results in general from material properties at a smaller scale, sometimes microscopic. However, in most cases, large scale properties are not the simple addition of small scale properties : one observes "collective" properties, which can be entirely new ones. In such a sense, some universal properties of disperse systems can be expec-

ted to exist. Therefore simple models for these general properties can
be developed in terms of a small number of variables and must be available for blood characterization. Conversely, although Blood and RBC suspensions appear as very complex systems, they present under normal conditions many advantages for the study of non-newtonian behaviour. Indeed,
the particles they contain have precise properties (in the sense of
small deviations from mean values) and are suspended in a fluid (the
plasma) having a well defined physico-chemical state. The most important
features are the following :

(i) the size of RBC corresponds to a very narrow calibration

(ii) their extreme deformability (however associated to a large resistan-
    ce to area changes) allows strong orientation effects by flow, dis-
    carding too high $\dot{\gamma}$ values to avoid any inertial effect or hemolysis.

(iii) Aggregates of RBC possess a characteristic shape, the so-called
    "rouleau-shape", and the aggregation mechanism does not seem to de-
    pend on the number of RBC in the rouleau, but mainly on RBC-RBC in-
    teraction, the latter varying with physico-chemical properties of
    suspending fluid.

(iv) Relaxation times associated with Brownian motion are very large and
    the observed characteristic time should be either the Maxwell rela-
    xation time of the collective (aggregated) structure or the one as-
    sociated with the single RBC deformation (and the resulting orienta-
    tion).

Clearly, much more studies are needed to establish unequivocally
the physiological significance of (i) the rheological properties of bulk
blood in vitro and blood flow in narrow vessels, (ii) the related models
and (iii) the variables involved in them.

Note added in proofs: A very complete information on hemorheology in diseases can be found in the recent book "Clinical Aspects of Blood Viscosity and Cell Deformability" - LOWE G.D.O., BARBENEL J.C. and FORBES C.D.
(Eds.) Springer-Verlag (Berlin), 1981.

REFERENCES

ANCZUROWSKI E., MASON S.G. (1967), The kinetics of flowing dispersions.
    III. Equilibrium orientations of rods and discs - *J.Colloid Interface
    Sci.* 23, 533-546.

ARRHENIUS S. (1917), The viscosity of solutions - *Biochem.J.* 11, 112-113.

BARBEE J.H. and COKELET G.R. (1971), The Farhaeus effect.- *Microvasc.Res.*
    3, 1-21.

BATCHELOR G.K. and GREEN J.T. (1972), The determination of the bulk stress
    in a suspension of spherical particles to order c^2 - *J.Fluid.Mech.*,
    56, 401-427.

BAYLISS L. (1952); Rheology of blood and lymph, *In Deformation and Flow
    in Biological Systems*, A. Frey-Wissling (Ed), North Holland Publ.,
    Amsterdam - Chap.6, pp 355-415.

BERGEL D.H. (1972), The rheology of human blood vessels - *In Biomechanics:
    its foundations and objectives* - Y.C. Fung, N.Perrone et M.Anliker
    (Eds) Prentice Hall, Inc.New-Jersey, pp.63-103.

BLOCH E.H. (1962) , A quantitative study of the hemodynamics in the living
    microvascular system - *Amer.J.Anatomy*,pp 110, 125-153.

BORN G.V.R., MELLING A. and WHITELAW J.H.(1978), Laser doppler microscope
    for blood velocity measurements - *Biorheology*, 15, 163-172.

BRINKMAN H.C. (1952), The viscosity of concentrated suspensions and solu-
    tions -*J.Chem.Phys.*, 20, 571.

BROCHARD F.(1977), Une bulle dégonflée: le globule rouge - *La Recherche*
    75, 174-177.

BROOKS D.E., GOODWIN J.W. and SEAMAN G.V.F. (1970), Interactions among
    erythrocytes under shear - *J.Appl.Physiol.*,28, 172-177.

BROOKS D.E. and SEAMAN G.V.F. (1971), Role of mutual cellular repulsions
    in the rheology of concentrated red blood cell suspensions - In *Theore-
    tical and Clinical Hemorheology*, H.H.Hertert and A.L. Copley (Eds),
    Springer Verlag, Berlin, pp 127-135.

BROOKS D.E., GOODWIN J.W., SEAMAN G.V.F. (1974), Rheology of erythrocyte
    suspensions: electrostatic factors in the dextran-mediated aggregation
    of erythrocytes - *Biorheology* 11, 69-77.

BROOKS D.E. (1976) Red cell interactions in low flow states - In *Micro-circulation*, vol.I, J.Grayson and W.Zingg (Eds), Plenum Press, New-York pp 33-52.

BUGLIARELLO G., KAPUR C., HSIAO G. (1965), The profile viscosity and others characteristics of blood flow in a non-uniform shear field - *Proc.Four.Int.Congr.on Rheology*, $\underline{4}$, A.L. Copley (Ed), Interscience, N.Y., pp 351-370.

CHIEN S., USAMI S., JAN K.M. (1971b), Fundamental determinants of blood viscosity - In *The Symposium on Flow*, Dowdell (Ed.), Pittsburg.

CHIEN S., LUSE S.A., JAN K.M., USAMI S., MILLER L.H. and FREMOUNT H. (1971c),Effects of macromolecules on the rheology and ultrastructure of red cell suspensions - In *Proc.6th Europ.Conf.on Microcirculation*, Karger, Basel (Eds.), 29-34.

CHIEN S., USAMI S., DELLENBACK R.J., BRYANT C.A. and GREGERSEN M.I.(1971d) Change of erythrocyte deformability during fixation in acetaldehyde - In *Theoretical and Clinical Hemorheology*, Hartett H.H. and Copley A.L. (Eds), Berlin, Springer-Varlag, pp 136-143.

CHIEN S.(1972), Present state of blood rheology - In *Hemodilution: Theoretical Basis and Clinical Applications*.Messmer K., Schmid-Schönbein H., S.Karger (Eds.), Basel, 1-40.

CHIEN S., KING R.G., SKALAK, R., USAMI S. and COPLEY A.L. (1975), Visco-elastic properties of human blood and red cell suspensions - *Biorheology*, $\underline{12}$, 341-346.

CHIEN S., (1979), Blood rheology - In *Quantitative Cardiovascular Studies* Hwang N.H.C., Gross D.R., Patel D.J. (Eds), Univ.Park Press, Baltimore (USA), 241-287.

COKELET G.R., MERRILL F.W., GILLILAND E.R. and SHIN H. (1963). The rheology of human blood measurement near and at zero shear rate - *Trans.Soc. Rheol.*,$\underline{7}$, 303-317.

COKELET G.R. (1972) Rheology of blood - In *Biomechanics, its Foundations and Perspectives*, Y.C.Fung, N.Perrone and M.Anliker (Eds), Prentice Hall, Inc., Englewood Cliffs, N.J., pp 63-103.

COKELET G.R. (1976) Macroscopic rheology and tube flow of human blood -
   In *Microcirculation* Vol 1.   J. Grayson and W.Zingg (Eds), Plenum
   Press. New York, pp 9-32

COPLEY A.L., HUANG C.R., KING R.G. (1973) Rheogoniometric studies of
   whole human blodd at shear rates from 1000 to 0.0009 sec^{-1}. Part 1.
   Experimental findings - *Biorheology* 10, 17-22. Part II. Mathematical
   interpretation. *Biorheology,* 10, 23-28.

COPLEY A.L., KING R.G., CHIEN S., USAMI S., SKALAK R. and HUANG C.R(1975)
   Microscopic observations of viscoelasticity of human blood in steady
   and oscillatory shear - *Biorheology* 12, 257-263.

COPLEY A.L., KING R.G., HUANG C.R.(1976) Erythrocyte sedimentation of
   human blood at varying shear rates - *Biorheology,* 13, 281-86.

COULTER Jr. N.A., MEGHA Singh (1971), Frequency dependence of blood viscosity in oscillatory flow - *Biorheology,* 8, 115-124

CROSS M.M. (1965) Rheology of Non-Newtonian fluids : A new flow equation
   for pseudoplastic systems - *Colloid Sci* 20, 417-437

DEVENDRAN T. and SCHMID-SCHONBEIN H.,(1975) Axial Concentration in Narrow
   Tube Flow for Various RBC Suspensions as Function of wall shear stress
   - *Pflügers Arch.* 355:R20

DINTENFASS L.,(1964), Rheology of the packed red blood cells containing
   haemoglobins AA, SA, SS - *J. Lab. Clin. Med.* 64, 594-603.

DINTENFASS L., BURNARD E.D. (1966a) Effect of hydrogen ion concentration
   on in vitro viscosity of packed red cells and blood at high hematocrits
   - *Med. J. Aust.* 1, 1072-1078.

DINTENFASS L., JULLIAN D.G. and MILLER G. (1966b), Viscosity of Blood in
   normal Subjects and in Patients Suffering from Coronary Occlusion and
   Arterial Thrombosis - *Am. Heart J.,* 71, 587-592.

DINTENFASS L., (1968), Internal viscosity of the red cell and a blood
   viscosity equation - *Nature, Lond.* 219, 956-957.

DINTENFASS L., (1969), The internal viscosity of the Red cell and the
   structure of the red membrane. Considerations of the liquid Crystalli-
   ne structure of the red cell interior and membrane from rheological
   data- *Mol.Cryst.* 8, 101-107.

DINTENFASS L. (1971) *Blood Microrheology Viscosity Factors in Blood Flow -Ischaemia and Thrombosis-* Butterworths. London.

DINTENFASS L. and KAMMER S. (1977) Plasma viscosity in 615 subjects. Effect of Fibrinogen, Globulin, and Cholesterol in Normals, Peripheral vascular.Disease Retinopathy and Melanoma - *Biorheology*, 14, 247-251.

DINTENFASS L. (1977) Blood Viscosity factors in severe non diabetic and diabetic retinopathy - *Biorheology*, 14, 151-157.

DINTENFASS L. (1979) Clinical applications of blood viscosity factors and functions : especially in the cardiovascular disorders - *Biorheology* 16, 69-84.

DORMANDY J.A. and EDELMAN J.B. (1973) High blood viscosity. An achological factor in deep venous thrombosis - *British Journal of Surgery*, 60, 187-189.

DUFAUX J., QUEMADA D., MILLS P. (1980) - Velocity profiles measurements by Laser-Doppler velocimetry (LDV) in plane capillaries. Comparison with theoretical profiles from a two fluid model. in : *Rheology*, Vol 3 G. Astarita, G. Marruci and L. Nicolais (eds) Plenum Press,NY;pp561-566

DIX F.J. and SCOTT-BLAIR G.W. (1940) On the flow of suspensions through narrow tubes - *J. Appl. Physics*. 11. 574-581.

FLAUD P., QUEMADA D.(1980) Rôle des effets non newtoniens dans l'écoulement pulsé d'un fluide dans un tuyau viscoelastique - *Revue Phys. Appl.* 15, 223-233

FISCHER Th. M., SCHMID-SCHONBEIN H., STOHR M.,(1978) Mechanical behaviour of human red blood cells in the shear field of viscous dextran solution - In *Cardiovascular and Pulmonary Dynamics*. Jaffrin M.Y.(ed),Editions Inserm Paris pp. 243-256.

FAHRAEUS R. (1929) The suspension stability of the blood - *Physiol. Rev.* 9. 241-274

FAHRAEUS R. and LINDQVIST T. (1931) The viscosity of the blood in narrow capillary tubes - *Amer. J. Physiol.* 96 : 562-568

GELIN L.E. (1961) Disturbances of the flow properties of blood and its counter action in surgery - *Acta Chirurgia, Scandinavia* , 122, 287-295.

De GENNES P.G. (1979). Conjectures on the transition from Poiseuille to plug flow in suspensions - *J. de Physique* 40, 783-787.

GILLESPIE T. (1963) The effect of Aggregation and Liquid Penetration on the viscosity of dilute suspensions of spherical particles - *J.Colloid Sci.* 18, 32-40.

GOLDSMITH H.L., MASON S.G. (1967) "The microrheology of dispersions" - In *Rheology: Theory and Applications.* Eirich, (Ed) Acad.Press, N.Y. pp. 85-250.

GOLDSMITH H.L., (1971) Deformation of human red cells in tube flow - *Biorheology* 7, 235-242.

GOLDSMITH H.L., (1968) The microrheology of red blood cell suspensions - *J. Gen. Physiol.* 52, 5s-28s.

GOLDSMITH H.L., (1973) The microrheology of human erythrocyte suspensions In *Proceed. XIIIe Int. Cong. Theor. and Appl. Mech.* E. Becker and G.K. Mikhailov (eds), Springer Verlag, Berlin, pp 85-103.

GREGERSEN, M.I., USAMI S., CHIEN S. and DELLENBACK R.J. (1967) Characteristics of torque-time records on heparinized and defibrinated elephant, human and goat blood at low shear rates ($0.01 \, sec^{-1}$) : effects of fibrinogen and Dextran (Dx 375) - *Bibl. anat.* 9 : 276-281.

HARKNESS W. (1971) The viscosity of human blood plasma. Its measurement in health and disease - *Biorheology* - 8, 171-193.

HAYNES R.H. and BURTON A.C. (1959) Role of the non Newtonian behavior of blood in hemodynamics. *Amer. J. Physiol.* 197-943.

HAYNES R.H. (1962) The viscosity of erythrocyte suspension - *Biophysics* 2 95-102

HEALY J.C., JOLY M.(1975) Rheological behaviour of blood in transient flow - *Biorheology* 12 335-340.

HOARE E.M., BARNES A.J. and DORMANDY J.A. (1976) Abnormal Blood Viscosity in Diabetes Mellitus and Retinopathy - *Biorheology*, 13, 21-25.

HOUWINK R. (1949) Macromolecular sols without electrolyte character. In: *Colloid Science, II, Reversible Systems,* 153. Kruyt H.R. (Ed) Elesevier Publ., Amsterdam.

HUANG C.R., SISKOVIC N., ROBERTSON R.W., FABISIAK W., SMITHERBERG E.H.,
COPLEY A.L. (1975), Quantitative characterization of whole human blood -
   Biorheology 12, 279-282.

ISOGAI Y., ICHIBIA K., IIDA A., CHIKATSU I. and ABE M. (1971), Viscosity
   of blood and plasma in various diseases - In *Theoretical and clinical
   hemorheology*, Hartett H.H. and Copley A.L. (Eds), Springer-Verlag,
   (Berlin) pp 136-143.

KARNIS A., GOLDSMITH H.L. and MASON S.G. (1966), The kinetics of flowing
   dispersions I: Concentrated suspensions of rigid particles. *J.Coll.
   Interface Sci.*, 22, 531-553.

KELLER J.B., RUBENFELD L.A. and MOLYNEUX J.E. (1967). Extremum principles
   for slow viscous flows with applications to suspensions - *J.Fluid.
   Mech.*, 30, 97-125.

KELLOG F. and GOODMAN J.R. (1960), Viscosity of blood myocardial infarction - *Circulation Research*, 8 , 972-978.

KLOSE H.J., VOLGER B., BRECHTELSBAUER H., HERNICH I. and SCHMID-SCHONBEIN
   H. (1972) - Microrheology and light transmission of blood I. The photometric quantification of red cell aggregation and red cell orientation - *Pflürers Arch.*, 333, 126-132.

KRIEGER I.M. and ELROD H. (1953) - Direct determination of the flow curves of non-newtonian fluids II: Shearing rate in the concentric cylinder viscometer - *J.Appl.Phys.*, 24, 134-140.

KRIEGER I.M. (in Surface and Coatings Related to Paper and Wood. R. Mar-
   chessault, C.Skaar ed. Syracuse Univ.Press (1967)) and T.J.DOUGHERTY,
   Some problems in the theory of colloids, (Ph.D.Thesis, Case Inst.Techn.
   (1959)).

KRIEGER I.M. (1963), A dimensional approach to colloid rheology - *Trans.
   Soc.Rheol.*, 7 , 101-109.

KRIEGER I.M. and DOUGHERTY T.J. (1959). A mechanism for non-newtonian
   flow in suspensions of rigid spheres - *Trans.Soc.Rheol.*, 3, 137-152.

LANDEL R.F., MOSER B.G. and BAUMAN A.J. (1965), Rheology of concentrated
   suspensions. Effect of a surfactant - In *Proceed. IVth Intern.Cong. on
   Rheology*, Part 2, Lee E.H. (Ed), Interscience, N.Y., pp 663-693.

LESSNER A., ZAHAVI J., SILBERBERG A., FREI E.H., and DREYFUS P. (1971) The viscoelastic properties of whole blood In *Theoretical and Clinical Hemorheology* ; H.H. Hartert and A.L. Copley (eds.) Springer-Verlag. New York, pp. 194-205

LINDSLEY H., TELLER D., NOONAN B., PETERSON M. and MANNIK M. (1973). Hyperviscosity Syndrome in Multiple Myeloma. A reversible, concentration dependent Aggregation of the Myeloma Protein - *The Amer of Medicine,* 54, 682-688.

MARON S.H. and SISKO A.W. (1957) Application of Ree-Eyring generalized flow theory to suspensions of spherical particles: II. Flow in low shear region - *J. Colloid.Sci.,* 12, 99-107.

MAUDE A.D. and WHITMORE R.L. (1958) Theory of the Blood Flow in Narrow Tubes - *J. Appl.Physiol.* 12: 105-113.

MATSUDA T. and MURAKAMI M. (1976). Relationship between fibrinogen and blood viscosity - *Thrombosis Research, Suppl.II,* 8, 25-33.

MERRILL I.W., MARGETTS W.G., COKELET G.R., BRITTEN A., SALZMAN E.W., PENNELL R.B. and MELIN M. (1955) Influence of plasma proteins on the rheology of human blood. In : *Proc.4th Inter.Cong.on Rheology.* A.L. Copley (ed.) Pt, 4, Interscience(Wiley), New York. pp 601-12.

MERRILL E.W., PELLETIER G.A. (1967) Viscosity of human blood : transition from newtonian to non-newtonian - *J. Appl. Physiol.* 23,178-182.

MERRILL E.W., (1969) Rheology of blood - *Physiol.Rev.* 49 : 863-888.

MIDDLEMAN S. (1972) *Transport phenomena in the cardiovascular system* Wiley-Interscience, N.Y. p. 91.

MILLER L.H., USAMI S. and CHIEN S. (1971) Alteration in the rheologic properties of Plasmodium Knowlesi - infected red cells. A possible mechanism for capillary obstruction - *J.Clin.Invest.* 50, 1451-1455.

MILLS P., QUEMADA D. and DUFAUX J. (1980) An optical method for studying RBC orientation and aggregation in a Couette flow of Blood Suspension. In: *Rheology* 3 G. Astarita, G. Marrucci, L. Nicolais (eds) Plenum Press NY 1980. pp. 567-572.

MOONEY M.(1951) The viscosity of a concentrated suspension of spherical particles - *J.Colloid Sci.* 6, 162-170.

MOORE F. (1959) (Cited by Cheng et Evans, 1965). *Trans.Brit.Ceram.Soc.*58, 470-492.

OSTWALD W., AUERBACH R. (1926) Uber die Viscosität kolloider Lösungen im Struktur, Laminar - und Turbulenzgebiet. *Kolloid z.* 38, 261-280.

PALMER A.A.(1968) Some aspects of plasma skimming. In:*Hemorheology* - A.L. Copley (ed.) Pergamon Press, Oxford. pp. 391-400.

PRAGER S. (1963). Diffusion and viscous flow in concentrated suspensions. *Physica* 29, 129-139 (1963).

QUEMADA D. (1976a). Red cell Aggregation and Thrombus formation: a rheological approach - *Proceedings of the 16th International Congress of Hematology: Topics in Hematology.* KYOTO, 1976. Excerpta Medica (Amsterdam) 415, 733-736.

QUEMADA D. (1976 b). Some new results in rheology of concentrated disperse systems and blood. In : *Proceedings of the VIIth International Congress on Rheology.* J. Kubat (ed.) Gothenburg, 1976, pp.628-629.

QUEMADA D. (1977) Rheology of concentrated disperse system and minimum energy dissipation principle. I. Viscosity-concentration relationship. *Rheol. Acta* 16, 82-94.

QUEMADA D. (1978 a) Rheology of concentrated disperse systems, II. A model for non newtonian shear viscosity in steady flows. *Rheol. Acta* 17, 632-642.

QUEMADA D. (1978b) Rheology of concentrated disperse systems. III. General features of the proposed non-newtonian model. Comparison with experimental data - *Rheol. Acta* 17, 643-653.

QUEMADA D., DUFAUX J., MILLS P.(1980) A two-fluid model for highly concentrated suspension flow through narrow tubes and slits: velocity profiles, apparent fluidity and wall layer thickness - In :*Rheology,* Vol.3 G.Astarita,G.Marrucci and L.Nicolais (eds) Plenum Press, NY pp.633-638

QUEMADA D. (1981) - A rheological model for studying the hematocrit dependence of red cell-red cell and red cell-protein interactions in blood *Biorheology* 18, 501-516 .

QUEMADA D., MILLS P., DUFAUX J., SNABRE P., LAMBERT M. (1981) Sedimentation effects in viscometric measurements - in:*Hemorheology and Diseases.*

J.F.Stoltz and P.Drouin(Eds)- Doin editeurs. Paris - pp 31-41.

REE T., EYRING H. (1955)- Theory of non-newtonian flow. I. Solid Plastic System. *J.Appl.Phys.*, 26, 793-804.

ROBINSON J.V. (1949) The viscosity of suspensions of spheres - J. Phys. and Colloid Chem. 53, 1042-1056.

ROSCOE R. (1952) The viscosity of suspensions of rigid spheres - *Brit. J. Appl. Phys.* 3, 267-269.

RUCKENSTEIN E. and MEWIS J. (1973) Kinetics of Structural Changes in Thixotropic Fluids - *Colloid and Interface Sci* 44, 532-541

SCHMID-SCHONBEIN H., WELLS R.E., GOLDSTONE J. (1971) Fluid drop-like behaviour of erythrocytes. Disturbance in pathology and its quantification *Biorheology* 7, 227-234.

SCHMID-SCHONBEIN H., WELLS R.E. (1971) Red cell aggregation and cell deformation : their influence on blood rheology in health and disease. In : *Theoretical and Clinical Hemorheology,* Hartet H.H. and Copley A.L. (eds) Berlin, Springer-Verlag, pp. 348-355.

SCHMID-SCHONBEIN H., GALLASCH G., VOLGER E., KLOSE H.J.(1973) Microrheology and protein chemistry of pathological red cell aggregation (blood sludge) studied in vitro - *Biorheology* 10, 213-227.

SCHMID-SCHONBEIN H. (1975) Erythrocyte rheology and the optimization of mass transport in the microcirculation - *Blood Cells* 2 285-306.

SCHMID-SCHONBEIN H. (1976) Microrheology of erythrocytes, blood viscosity and the distribution of blood flow in the microcirculation. In : *International Review of Physiology. Cardiovascular Physiology* - A.C. Guyton and A.W. Cowley (eds) University Park Press. Baltimore pp. 1-62

SCHMID-SCHONBEIN H., FISCHER T., DRIESSEN G., RIEGER H. - Microcirculation In : *Quantitative Cardiovascular Studies: Clinical and Research Applications of Engineering Principles,* N.H.C. Hwang, D.R. Gross and D.J. Patel (eds) Univ. Park Press, Baltimore (1979), Chap.8, 353.

SCHOLZ P.M., KARIS J.H., GUMP F.E., KINNEY J.M. and CHIEN S. (1975) Correlation of blood rheology with vascular resistance in critically ill patients - *J.appl.Physiol.* 39 1008-1011

SCOTT BLAIR G.W.(1959) An equation for the flow of blood, plasma and se-
    rum through glass capillaries - *Nature* 183, 613-615.

SEGRE G. and SILBERBERG A. (1962) Behavior of macroscopic rigid spheres
    in Poiseuille flow. *J. Fluid Mech.* 14, 115-135 : 136-157

SUTERA S.P. (1978) Red cell motion and deformation in the microcircula-
    tion - In : *Cardiovascular and Pulmonary Dynamics*. Jaffrin M.Y.ed. Edi-
    tions,Inserm Paris pp.221-242.

TAYLOR G1 (1932) The viscosity of a fluid containing small drops of ano-
    ther fluid - *Proc.Roy.Soc.(London)* 138A 41-45

THOMAS H.W. (1963) The Wall Effect in Capillary Instruménts, *Biorheology*
    1 : 41-56.

THURSTON G.B. (1976) The viscosity and viscoelasticity of blood in small
    diameter tubes - *Microvasc. Res.* 11:133-146

THURSTON G.B. (1979a) Erythrocyte Rigidity as a Factor in Blood Rheolo-
    gy : Viscoelastic Dilatancy - *J. of Rheology*, 23, 703-719.

THURSTON G.B. (1979b) Rheological parameters for the viscosity viscoelas-
    ticity and thixotropy of blood - *Biorheology*, 16, 149-162.

USAMI S., CHIEN S. and GREGERSEN M.I.(1971) - Viscometric Behavior of
    Young and Aged Erythrocytes - In *Theoretical and Clinical Hemorheolo-
    gy*, Hartett H.H. and Copley A.L.(eds) Springer-Verlag, Berlin,136-143.

VAND V. (1948) Viscosities of solutions and suspensions - *J. Phys.Coll.*
    Chem. 52, 277-299.

VINCENT N.M., OLIVER D.R.(1977) Blood sedimentation at controlled shear
    rates - *Biorheology* , 14, 51-58.

WEINBERGER C.B. and GODDARD J.D. (1974) Extensional flow behaviour of
    Polymer solutions and particle suspensions in a spinning motion - In-
    tern. *J. Multiphase Flow*, 1, 465-486.

WHITMORE R.L. (1967) - A theory of blood flow in small vessels - *J. Appl.*
    Physiol., 22, 767-771.

... ...

CHAPTER II

THE ARTERIAL WALL

J.C. Barbenel,
Bioengineering Unit
University of Strathclyde,
Wolfson Centre,
Rottenrow
GLASGOW G4 ONW. U.K.

1.   THE GEOMETRY AND STRUCTURE OF ARTERIES

1.1.   Introduction

The pioneer of the quantitative study of the flow of blood through tubes was Poissuille, who was both physicist and physician. He modelled the flow of blood through the circulation by investigating the flow of water through rigid cylindrical tubes. He was forced to use water as he was unable to prevent the blood from clotting, and he was fortunate in his substitution because blood shows non-Newtonian behaviour, which will be dealt with elsewhere in this volume. He was also fortunate in using rigid tubes. The real blood vessels have complicated non-cylindrical geometries, and are highly extensible and non-linear. If he had used blood and realistic models of blood vessels, it is unlikely he would have produced the clear cut results he did.

This section will discuss the arterial wall, both geometry and mechanical properties. The data from different sources gives values of parameters, often markedly different and often conflicting. There are several reasons for this.

Results have been obtained from measurements made both in living and dead animals. The blood vessels are highly deformable and flaccid when empty. The reported values of thickness, diameter, length, etc., may be open to some doubt as the methods used to measure these variables often lead to changes in dimensions due to forces applied by the measuring instruments. During life the large vessels are under a state of longitudinal tension, and when cut transversely will retract. Further retraction will occur if they are dissected free of the surrounding tissue (McDonald, 1974.[1]). Thus data obtained on excised specimens may be distorted by the loss of this resting tension. Similarly during life the vessels are filled with blood at a varying pressure, and it is often difficult to decide what physiological significance some of the *in vitro* data has. Finally there is usually a very considerable range of values for the same feature of structure measured in different animals.

1.2.   The Size of Arteries

The circulation within the arterial system is usually discussed in terms of large and small arteries and arterioles. The dividing line is usually arbitrarily set at vessels having a diameter of about 0.1 mm.

The difference in size also reflects important differences in flow
conditions. The relative importance of the inertial forces can be
expressed in the Reynold's Number Re;

$$Re \ = \ \frac{\bar{U}d}{\nu} \tag{1}$$

Where $\bar{U}$ is the mean flow velocity,

       d the diameter of the vessel

and ν the kinematic viscosity.

In the larger vessels Re is large and inertial forces more important than
the viscosity. In vessels of less than 0.1 mm diameter Reynold's number
is typically less than unity, and viscous forces become increasingly
important. The red blood cell has a diameter of 7-8 μm and the 0.1 mm
diameter is also a point at which the structure of the blood becomes
important, and it becomes less realistic to treat the blood as a continuum.

The small vessels are extremely important, but most mechanical
testing has been carried out on the larger vessels and the results
extrapolated to the smaller structures.

1.3.   The Geometry of the Large Vessels

The aorta is the major artery carrying the blood from the left
ventricle. The vessel is curved three dimensionally, first running in
the direction of the head, where it gives rise to branches supplying the
head and upper limbs. The aorta then curves to run caudally (Figure 1.1).
After this follows a nearly straight course, giving rise to arteries
supplying the organs in the abdominal cavity, and terminates by dividing

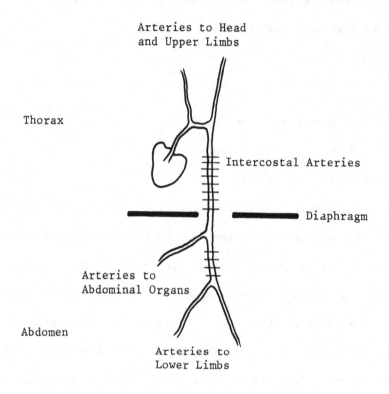

Figure 1.1   Schematic diagram of aorta

into two major vessels supplying the lower limbs. In the dog, and
similar animals, a third terminal branch runs to the tail.

Most of the detailed information on artery size, diameter etc.,
has been derived from dogs. This is the most reliable data, but great
care must be used in attempting to apply or extrapolate the results to man.

The aorta tapers along its length, and in dogs it can be described by the exponential equation

$$A = A_o \, e^{-(Bx/R_o)}$$ (2)

A is the aortic area at length x from the start of the vessel where the area and radius is A_o and R_o. B is a parameter, the value of which lies between $2 - 5 \times 10^{-5}$ (Caro et al, 1978[2]). Details of the cross sectional area of the arteries of dogs, at selected sites, will be found in McDonald, 1974.

1.4. The Area of Branches and Branching Angles

The cross sectional area of the aorta decreases regularly with the distance from the heart, but because of the presence of branches, the total arterial cross section actually increases. There is a considerable body of data on the cross sectional area of dog arteries obtained from casts made by expanding the arteries with elastomers. This is discussed at length by Iberall, 1967[3].

The angle between the arterial branches and the vessel from which they arise is usually called the branching angle. There is a very wide range of angles found in the circulation, but a lack of detailed information. It has been known, however, that there is general relationship between the branching angle and the relative size of branches (d'Arcy Thompson, 1945[4]):

(i) if the artery branches into two equal branches (e.g. the common iliac arteries) the branching angles are symmetrical.

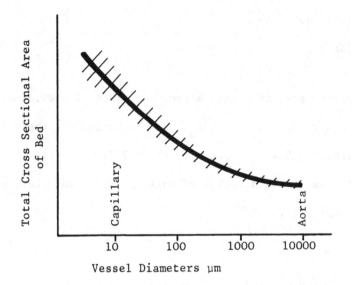

Figure 1.2 Cross sectional area of arterial tree (After Iberall, 1967)

(ii)   if the branches are of unequal size the smaller vessel has a larger branching angle (e.g. internal and external iliac arteries).

(iii) branches very much smaller than the parent artery have a branching angle of about 90° (e.g. posterior intercostal arteries).

1.5   The Structure of the Arterial Wall

The walls of arteries are neither uniform nor homogeneous but consist of a variety of cells and extracellular proteins. The arrangement of these components and the structure of the arterial wall is described in the standard text books of histology (e.g. Bloom and

Fawcett, 1975[5]), and a detailed description will be found in Benninghoff (1930[6]).

The dominant structural materials are the fibrous proteins, collagen and elastin, and the associated glycosominoglycan ground substance. The major cellular component is smooth muscle. The organisation of these components in the blood vessel wall is similar in all arteries and they are arranged in three concentric layers - the tunica intima, which is the innermost layer and surrounded by the tunica media and surrounding both is the tunica adventitia. Each layer is separated by a layer of elastin, the internal elastic lamina demarcating the intima-media junction and the external elastic lamina marking the junction between the media and adventitia. In each of the three layers there is a dominant structure and cell type.

The tunica intima consists of a single, thin (0.5 μm) layer of endothelial cells which forms a continuous lining to the circulatory system. The cells are mechanically fragile but damaged or lost cells can be replaced by regeneration. Beneath the endothelial cells there is a thin layer of collagen fibres.

The internal elastic lamina beneath the adventia consists of a dense layer of elastic fibres, which has also been described as an elastic membrane.

The tunica media is usually the thickest layer of the arterial wall,

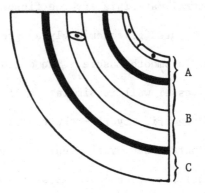

Figure 1.3 Diagramatic cross section of the arterial wall

A = Tunica Intima, with a layer of endothelial cells
    and connective tissue.

B = Tunica Media, showing elastic laminae connected by
    oblique smooth muscle cell.

C = Tunica Adventia.

and contains elastin and collagen fibres, and smooth muscle cells. There

is a considerable variation in the amount of elastin and muscle in

different arteries and this leads to the description of arteries as

being either elastic or muscular. There is no sharp demarcation between

the two types of arteries but the major vessels nearer the heart are predominantly elastic and the distant arteries muscular.

The tunica media of a typical elastic artery consists of multiple concentric layers of elastin separated by thin layers of collagen and smooth muscle cells. The thickness of these layers of elastin is relatively constant and they appear in the large arteries of all mammals (Wolinsky and Glagov, 1967.[7]). Thus the number of layers is approximately proportional to the wall thickness. The elastic fibres in the layers are arranged in a spiral pattern.

In the more distal arteries the media consists mainly of spirally arranged smooth muscle cells, which are arranged in concentric layers separated by small amounts of collagen and elastin. The number of layers of muscles diminishes with the radius of the vessel.

The external elastic lamina is a thin layer of elastin, less clearly defined than the internal lamina.

The tunica adventitia consists of loose connective tissues containing both collagen and elastin fibres which are in a predominantly longitudinal direction. The outer demarcation is not clearly defined and merges with the tissue surrounding the vessel.

The nourishment of the intima and inner layers of the media is derived from the blood flowing through the vessels, but arteries greater than 1 mm in diameter have small vessels called vasa vasorum supplying the vessel wall.

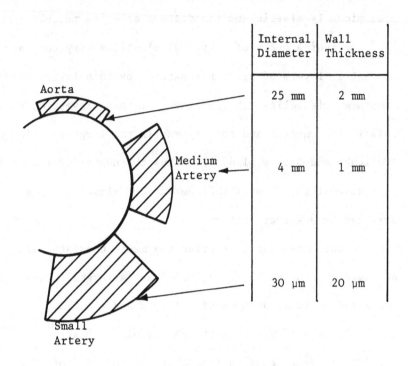

	Internal Diameter	Wall Thickness
Aorta	25 mm	2 mm
Medium Artery	4 mm	1 mm
Small Artery	30 μm	20 μm

Figure 1.4   Relative thickness of relaxed arterial wall
              (After Burton, 1962[8])

              Equivalent values for vessels in which the muscle of
              the tunica media is contracted will be found in
              Folkow and Neil (1971[9]).

1.6. Thickness of the Arterial Wall

    The wall thickness of the larger arteries varies with the size of

the animal. The results of Wolinsky and Glagov (1967) indicate a rapid

increase in thickness with body weight for small animals, but for large

animals, those which weigh more than 60 Kg, the thickness is relatively

constant. The diameter of the arterials also changes in a parallel manner, and thickness to diameter ratios appear to be relatively constant, being $\simeq$ 0.1.

In the smaller peripheral arteries the wall becomes thinner, but the thickness to diameter ratio actually increases, and for arterioles the ratio may be 0.4.

The size of the arteries also change with age, and during growth both wall thickness and diameter increase, but the ratio remains relatively constant (Caro et al, 1978). The increase in thickness of the wall is due to thickening of and an increase in number of the elastic lamellae of the tunica media. After maturity has been reached there is a continuing increase in diameter and wall thickness (Learoyd and Taylor, 1966[10]).

2.   EXPERIMENTAL MECHANICS OF THE ARTERIAL WALL

There is considerable disagreement on how the properties of the arterial wall should be analysed. Before discussing these areas of uncertainty this section will review, in descriptive terms, what is the generally accepted view of the mechanics of the arterial wall.

The mechanical properties of the arterial wall have been investigated both in quasistatic and dynamic test modes. Tests have been made on arterial segments *in vivo* - that is with the specimen still in place in the living experimental animal but the majority of results have been obtained *in vitro* on excised arteries. These excised arteries may be maintained at the *in vivo* length or allowed to relax before testing. In

addition there are numerous results which have been obtained from strips
on rings cut from the wall of excised vessels.

The results of these tests may differ in the magnitude of the
descriptive parameters obtained, but the general form of the load-
deformation response of the walls of the larger vessels have been clearly
established. Not unexpectedly, the response is similar to other soft
connective tissues and this reflects the fact that the behaviour of these
tissues is controlled by the properties, arrangement and mechanical
properties of the collagen and elastic fibres, and ground substance
common to all these tissues.

2.1.   Quasistatic load-extension response

The load-extension response obtained from excised strips and the
equivalent pressure-radius results obtained from inflated arterial
segments are all markedly non-linear. There is an initial phase during
which small load increments produce large increases in deformation
(Figure 2.1).   The vessel becomes stiffer as the load (or deformation)
increases. The stiffness increases monotonically, although at an ever
decreasing rate. As well as being non-linear the vessels are non-elastic.
When the specimen is unloaded, load-deformation curve does not coincide
with that obtained for increasing load.

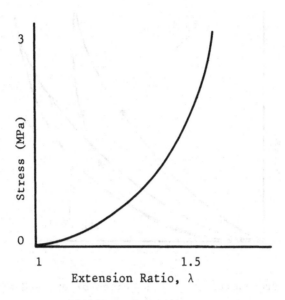

Figure 2.1 Typical load-deformation curve for excised artery.
Uniaxial Tension.

The vessel response also shows preconditioning, a progressive change
in response produced by repeated load or deformation cycling (Figure 2.2).
Typically the curvature of the load-deformation curve increases and the
magnitude of the hysteresis curve decreases as the cycles are repeated.
The alteration in behaviour grows progressively smaller as the number of
cycles increase, and after a relatively small number of cycles (usually
less than ten) a stable and repeatable response is obtained. As Fung
(1981[11]) has pointed out, the specimen is preconditioned only for the
specific test cycle used. If this is altered the specimen must be
preconditioned anew using the required test cycle. The majority of

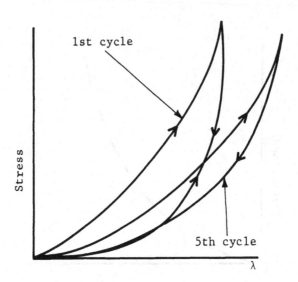

Figure 2.2   Preconditioning in specimen of excised artery.
Uniaxial Tension.

experimental results which are reported in the literature has been

obtained on preconditioned tissue.

2.2.   Time Dependence

Hysteresis is typical of materials which show time dependence, and

other aspects of this behaviour can be shown in arteries. An important

exception is the insensitivity of load-extension and hysteresis behaviour

to strain rate. Tests carried out in the range 10^{-3} - 1 s^{-1} show little

change in response (Tanaka and Fung, 1974[12]), and rate dependence appears

only at high rates (Collins ánd Hu, 1972[13]). This feature is found in

other soft connective tissues.

Stress Relaxation at constant strain has been reported, mainly in experiments made with strips and rings of tissue. The response is typical of non-linear viscoelastic materials, with greater relaxation occurring at higher strains (Figure 2.3).

The amount of relaxation is also site dependent, with muscular arteries showing greater relaxation than elastic vessels.

The relaxation is most rapid at short times, and the rate of relaxation decreases monotonically with time. If the stress is plotted against the logarithm of elapsed time, an almost linear relationship is usually obtained.

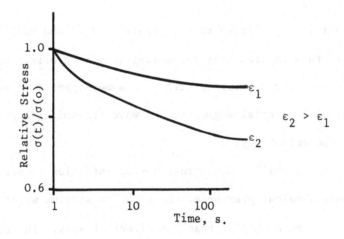

Figure 2.3   Stress relaxation in excised artery

Creep behaviour has also been reported both in excised strips and in pressurised arterial segments (e.g. Wiederhielm, 1965[14]). As with stress relaxation, the response is non-linear and site dependent.

Dynamic behaviour can be characterised in terms of the storage
modulus and phase angle. There is general agreement that the storage
modulus shows little frequency dependence above about 2Hz, but it
decreases slowly down to 10^{-3}Hz. The ratio of the storage modulus to the
quasistatic value is site dependent. The ratio is higher (1.6-2) in the
peripheral muscular arteries, than in the central elastic vessels where
the ratio is 1.1.

The phase angle, which is commonly about 10°, also shows little
variation with frequency.

2.3.   Anisotropy

When arteries are inflated they elongate, and if the wall is
incompressible, this implies that the vessel is anisotropic. Load-
elongation measurements on excised strips, the measurement of wall strains
during inflation of arterial segments, and wave transmission studies, have
all confirmed the anisotropy.

Patel and Fry (1969[15]) showed that during inflation of arterial
segments to physiological pressures, the shearing strains which were
produced were very much smaller than the direct strains. The results
suggest that the wall shows elastic symmetry and that it is cylindrically
orthotropic. In these circumstances, the number of elastic constants
necessary to describe a compressible material is reduced to nine, which
may be taken as three orthogonal tensile moditi, three Poisson's ratios
and three shear moduli. If the wall material is incompressible, the
number is further reduced (Patel and Vaishnav, 1980[16]).

The results have been confirmed by other workers and by other means, and there is general agreement that the vessel walls show cylindrical orthotropy. How to treat the walls quantitatively is less generally agreed.

The results of Tanaka and Fung (1974), indicate the existence of anisotropic time dependence. Strips of artery cut longitudinally showed greater, and more rapid, relaxation than strips cut circumferentially.

2.4.   Compressibility of the Arterial Wall

The arterial wall has a high water content and it has been thought that it is incompressible. Lawton (1954[17]) stretched strips of aorta in saline and showed only a small change in volume, and concluded that the aortic wall was essentially incompressible. Similar experiments by Carew et al (1968[18]) on inflated and stretched arterial segments showed that the relative volume change on straining was less than 1%, and confirmed the incompressibility of the tissue. Both experiments would fail to demonstrate a volume change due to loss of fluid from the arterial wall. Bergel (1972[19]) suggests such a mobile component of the water within the wall is unlikely. Such a change in volume due to fluid loss does, however, appear to occur during *in vitro* tests on skin, and has been reported for cat's skin (Veronda and Westmann, 1970[20]), also for human skin (Stark and Al Haboubi, 1981[21]).

Tickner and Sacks (1967[22]) made radiographic estimations of the alteration in wall thickness during simultaneous inflation and elongation of human and canine arteries and reported a change in volume. For the

volume to remain unaltered, the sum of the Poisson's ratios of the
material must be 1.5, and the data of Tickner and Sacks show that this is
not the case. The use of radiography to determine wall thickness may
lead to considerable errors. It has also been suggested that the results
were due to water loss and drying during the inflation process. In order
for the volume to decrease, the sum of two of the Poisson's ratios should
exceed unity. This certainly appears to occur in some tests (Tickner
and Sacks, Figure 7a) but is not a general result.

Although the evidence for arterial incompressibility may be less
convincing than much of the literature suggests, the wall can usefully be
considered incompressible.

2.5.   Structural Basis of Mechanical Response

There has been considerable interest in interpreting the overall
mechanical properties of the soft connective tissues in terms of the
arrangement and properties of their constituents. The major structural
components of the arterial wall are the elastin and collagen fibres and
muscle cells found in the tunica media. At physiological distending
pressures, the components are well oriented and form a series of well
defined layers (Wolinsky and Glagov, 1964[23]).

Elastin has been extracted from ligamentum muscle and tested in
uniaxial tension (Carton et al, 1962[24]); the results showed considerable
non-linearity.

More recent work (Jenkins and Little, 1974[25]) showed the material to
be more linear with a tensile modulus of c.0.5 M Pa. All results show,

however, that the material is highly extensible and almost elastic, with time dependence being small (Gosline, 1976[26]).

Collagen occurs as crimped fibres, and during fibre strengthening the fibres may undergo considerable extensions. The nature of the crimp, and therefore the magnitude of the extension is widely variable.

The straightened collagen fibres are stiffer than elastin, with a tensile modulus about two orders of magnitude greater (Bergel, 1972; Haut and Little, 1972[27]). The fibres show considerable time dependence.

There is considerable disagreement about the mechanical properties of the muscle cells in the wall, but it appears that the active contractile force produced by the cells may be equivalent to a modulus of the order of 0.1 M Pa.

Studies (Dobrin and Rovick, 1969[28]) of the static properties of elastic arteries in which the muscle tone was abolished by potassium cyanide or heightened by nor-ephinerphrine showed the muscle to have a significant effect on the circumferential modulus. There was almost no effect on the longitudinal properties.

Between the muscle cells and the elastin and collagen fibres, there is abundant glycosaminoglycen ground substance. This material is associated with a large amount of fluid, and up to 70%, by weight, of the vessel wall can be made up of water (Harkness, et al 1957[29]). It is not clear whether any of this water is mobile. If such a mobile component exists it may be unable to take part in the load transmission mechanisms of the arterial wall, and may be extruded on straining, thus making the wall effectively compressible. Recent studies (McCrum and Dorrington, 1976[30]) suggest that when elastin is stretched, the material absorbs water, which is then desorbed on releasing the fibres. This implies that there must be some mobile water present. The authors suggest that this also means that elastin must be considered to be compressible.

There is a considerable site variation of the ratio of elastin to collagen, (McDonald, 1974). In the thoracic aorta there is about twice as much elastin than collagen. At the diaphragm, this falls so that there is about three times as much collagen as elastin, and this ratio persists in the muscular arteries.

The geometry and properties of the elastin and collagen fibres has led to the suggestion that the former is stretched during the initial phase of extension of the wall and that during this period, the crimps are lost from the collagen which is important at the stiffer, high extension phase. Evidence to support this has been obtained from studies in which either component was enzymatically degraded (Roach and Burton, 1957[31], Hoffman et al, 1973[32]).

3.   STRESS-STRAIN RELATIONS

The walls of the arteries undergo large deformations, showing a non-linear relationship between stress and strain, and are time dependent. Most studies of the stress-strain response of blood vessels have assumed that the effect of time dependence can be rendered insignificant if suitable test methods are used. The favourite methods are the application of constant loading or deformation rates (e.g. Tanaka and Fung, 1974) or incremental measurements made after the majority of the creep or relaxation has occurred (e.g. Patel et al, 1969[33]). It is usual in

these conditions to talk of the elastic properties of the arterial wall.

The problem of the large strain non-linearity can be tackled by the

use of a suitable non-linear strain measure. Fung(1972[34]) showed that

the gradients of the load (T)/extension (λ) plots obtained by constant

strain rate tests were directly proportional to the load,

$$\frac{dT}{d\lambda} = \alpha(T + \beta) \qquad \alpha,\beta \text{ constants}$$

$$\text{hence } T = ce^{\alpha\lambda} - \beta \tag{3}$$

The exponential formulation was incorporated into a stored energy function

to describe arterial elasticity by Vito, (1973[35]). A more detailed

study by Tanaka and Fung (1974) showed that the linearity was confined

to the large strain region, and did not describe the total stress-strain

curve.

There are two theoretical approaches by which general constitutive

relationships may be derived.

The pressure of blood in an artery during function is such that the

vessel normally operates in an initially strained state, with extensions

being as large as 70%. The strain variations about this resting state

are very much smaller (Patel et al, 1969). This suggests that

constitutive relations be sought in terms of incremental stress-strain

relations (Biot, 1965[36]). The suitability of the method for application

to normal function is self-evident. There are, however, major

disadvantages if a wide range of strains are to be considered, as the

response around a large number of average strains will require to be
investigated. It may also be difficult to compare incremental moduli
obtained under different conditions.

The alternative approach in which general three dimensional stress-
strain relations are evaluated has relied heavily on the stored energy
function formulation which has been applied to elastomers. The theory
is very general, and allows the calculation of incremental moduli if
desired. It is, however, of considerably greater complexity than the
incremental approach.

3.1.   Incremental Elasticity

Bergel (1961[37])investigated the variation of the radius of excised
dog artery subjected to different internal pressures. The vessel was
maintained at its *in vivo* length.

The vessel was assumed to be isotropic and incompressible. The
Young's modulus (E) of such a tube of fixed length is given by

$$E = \frac{\Delta p \; 2(1-\nu^2)R_i^2 \; R_o}{\Delta R_o \; (R_o^2 - R_1^2)} \tag{4}$$

where Δp is the change in internal pressure, ΔR_o the change in external
radius R_o, R_i the internal radius and ν Poisson's ratio, which was
taken as 0.5 (see Bergel 1972 for derivation).

The relationship in the above equation was adapted to provide an
incremental modulus E_{inc} at a pressure p_2 (with radii Ro_2 and Ri_2) from
measurements made at a lower pressure p_1 (external radius Ro_1) and at a
higher pressure p_3 (external radius Ro_3) when

$$(E_{inc})_{p2} = \frac{(P_3 - P_1) \, 2(1 - 0.5^2)R_{i2}^2 \, R_{o2}}{(R_{o3} - R_{o1})(R_o^2 - R_i^2)} \qquad (5)$$

The results indicate that the magnitude of the incremental modulus was an increasing function of the internal pressure. The modulus increased approximately fifteenfold with an increase of pressure from 40 to 220 mm Hg(5.2 - 29k Pa).

The major limitation of the analysis was the assumption of isotropy. Even if the vessel was isotropic in the resting state, the strains produced by stretching it to the *in vivo* resting length would produce unequal strains in the three principal directions and, therefore, anisotropy.

The evaluation of the anisotropic incremental moduli for the aorta was made by Patel et al (1969). The wall was once again treated as incompressible but orthotropic. Tests were made on dogs *in vivo*, with *in vitro* comparisons corrected for arterial tethering affects (Patel and Fry, 1966[38]).

The vessel was considered to be a thin walled cylinder and described in cylindrical co-ordinates such that the z co-ordinate corresponded to the centre line of the aorta, the r co-ordinate to the radial direction and θ to the circumferential direction. For an arterial segment of length L, midwall radius R and thickness h subjected to an internal pressure p and longitudinal tension f the incremental strains in these three co-ordinate directions will be given by

$$e_r = \frac{\Delta h}{h_o} = \frac{h - h_o}{h} \qquad (6a)$$

$$e_\theta = \frac{\Delta R_o}{R_o} = \frac{R - R_o}{R_o} \qquad (6b)$$

$$e_z = \frac{\Delta L}{L_o} = \frac{L - L_o}{L} \tag{6c}$$

The stresses in these three directions were:

$$\left.\begin{array}{l} \sigma_r = -p/2 \\[2mm] \sigma_\theta = pR/h \\[2mm] \sigma_z = f/2\pi Rh \end{array}\right\} \tag{7}$$

The equivalent incremental stresses were denoted by P_r, P_θ and P_z.
The incremental normal strains and stresses are related by the

matrix equation

$$\begin{bmatrix} e_r \\ e_\theta \\ e_z \end{bmatrix} = \begin{bmatrix} C_{rr} & -C_{r\theta} & -C_{rz} \\ -C_{\theta r} & C_{\theta\theta} & -C_{\theta z} \\ C_{zr} & -C_{z\theta} & C_{zz} \end{bmatrix} \begin{bmatrix} P_r \\ P_\theta \\ P_z \end{bmatrix} \tag{8}$$

with Cij = Cji. The assumption of incompressibility requires that
$e_r + e_\theta + e_z = 0$. The constants Cij are related to the incremental
Young's moduli (E_θ, E_z and E_r) and Poisson's ratios ($\nu_{\theta z}$, ν_{zr}, $\nu_{r\theta}$) by:

$$E_\theta = \frac{1}{C_{\theta\theta}} \; ; \; E_z = \frac{1}{C_{zz}} \; ; \; E_r = \frac{1}{C_{rr}} \tag{9}$$

and

$$\left.\begin{array}{l} C_{z\theta} = \dfrac{\nu_{z\theta}}{E_\theta} = \dfrac{\nu_{\theta z}}{E_z} \\[4mm] C_{zr} = \dfrac{\nu_{zr}}{E_r} = \dfrac{\nu_{rz}}{E_z} \\[4mm] C_{r\theta} = \dfrac{\nu_{r\theta}}{E_\theta} = \dfrac{\nu_{\theta r}}{E_r} \end{array}\right\} \tag{10}$$

The values of the Young's moduli were evaluated for the thoracic
aorta of dogs. The vessel was maintained at one of three chosen initial

lengths and the internal pressure increased in incremental steps of
20 cm H_2O (1.9k Pa). Each was maintained for 1 minute to allow creep,
and the radius and longitudinal load measured.

The pressure, radius, and longitudinal loads were used to calculate
the incremental strains and stresses. The incremental stress may be
calculated directly from the measured variables (Fung, 1977[31]) but were
calculated by Patel et al from the stresses calculated, using equations
6 & 7 for each state of strain.

The results indicated that the incremental stresses and strains
were small compared to the average strains, thus justifying the use of
incremental theory.

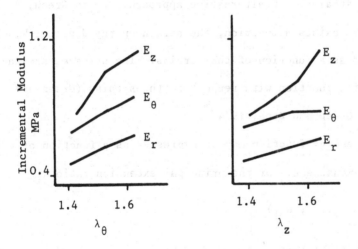

Figure 3.1   Variation of incremental moduli with extension ratio.

The magnitude of the moduli increased with the extension ratios λ_θ and λ_z as would be expected for a non-linear material. In general $E_z > E_\theta > E_r$ at physiological pressures.

The vessel walls are clearly anisotropic. The errors produced by treating an orthotropic tube as isotropic were analysed by Hardung (1964[40]) and Bergel and Schultz (1971[41]) used the results to show that there was good agreement between the results of Bergel (1961)   and Patel et al (1968).

3.2.   Strain Energy Density or Stored Energy Function

The use of the incremental stresses and strains follows the Cauchy approach to elasticity in which the stresses are expressed directly as a function of the strains. An alternative approach, due to Green, is to assume that there exists a function, the strain energy density, W, which may be expressed as a function of the strains. The stresses are the derivatives of the function with respect to the strains (Green and Adkins, 1960[42], Green and Zena, 1954[43]).

The strain energy function may be expressed as a function of a suitable strain measure ε_{ij} or the principal extension ratio λ_i

$$
\left.
\begin{aligned}
W &= W(\varepsilon_{11}, \varepsilon_{22}, \ldots, \varepsilon_{31}) \\
W &= W(\lambda_1, \lambda_2, \ldots, \varepsilon_{31})
\end{aligned}
\right\}
\tag{11}
$$

For isotropic materials the strain dependence may be replaced by a dependence on the principal strain invariants

$$
W = W(I_1, I_2, I_3)
\tag{12}
$$

where

$$I_1 = \lambda_1^2 + \lambda_2^2 + \lambda_3^2$$

$$I_2 = (\lambda_1\lambda_2)^2 + (\lambda_2\lambda_3)^2 + (\lambda_3\lambda_1)^2 \qquad\qquad (13)$$

$$I_3 = (\lambda_1\lambda_2\lambda_3)^2$$

The principal stress-strain relations are obtained from

$$\sigma_i = \frac{\partial W}{\partial \lambda i}$$

$$= \frac{\partial W}{\partial I_1}\frac{\partial I_1}{\partial \lambda_i} + \frac{\partial W}{\partial I_2}\frac{\partial I_2}{\partial \lambda_i} + \frac{\partial W}{\partial I_3}\frac{\partial I_3}{\partial \lambda_i} \qquad\qquad (14)$$

Stress-strain relations for specific deformations may be obtained by manipulation of equations 14.

The third invariant I_3 is the relative change in volume. For incompressible materials the term $\frac{\partial W}{\partial I_3}$ is undefined. As a result the stresses are indeterminant, and can be obtained only with respect to an arbitrary hydrostatic pressure.

The application of this version of the theory to the arterial wall is discussed by Tickner and Sacks (1964), but as we have seen the vessel wall is not isotropic. More realistic models of wall anisotropy produce considerably more complexity.

Vaishnav et al (1972[44]) investigated the strain energy density of the thoracic aorta of dogs on the assumption, used in the incremental study (Patel et al 1969, See Section 3.1) that it was incompressible and orthotropic. For such a material the strain energy density may be expressed as (Green and Adkin, 1960, p.28)

$$W = W(\varepsilon_{11}, \ \varepsilon_{22}, \ \varepsilon_{33}, \ \varepsilon_{12}^2, \ \varepsilon_{23}^2, \ \varepsilon_{31}^2) \qquad\qquad (15)$$

Vaishnav et al used the Green strain components which, in the notation used for the incremental study, are

$$\epsilon_r = \tfrac{1}{2}(\lambda_r^2 - 1); \quad \epsilon_\theta = \tfrac{1}{2}(\lambda_\theta^2 - 1); \quad \epsilon_z = \tfrac{1}{2}(\lambda_z^2 - 1) \tag{16}$$

It was assumed that W could be approximated by a polynomial of the form

$$W = k_1\epsilon_r^2 + k_2\epsilon_\theta^2 + k_3\epsilon_z^2 + k_4\epsilon_{r\theta}^2 + k_5\epsilon_{\theta z}^2 + k_6\epsilon_{zr}^2 + k_7\epsilon_r\epsilon_\theta + \ldots . \tag{17}$$

The special case of the simultaneous inflation and longitudinal extension was considered. Under such tests the principal strains are very much smaller than the shear strains and $\epsilon_{r\theta}$, $\epsilon_{\theta z}$ and ϵ_{zr} were put equal to zero. The incompressibility condition meant that the three principal strains were interrelated and the polynomial series in equation 11 was approximated by finite series in ϵ_θ and ϵ_z. Three series of second, third and fourth degree were proposed. These are respectively

$$W = A_1\epsilon_\theta^2 + B_1\epsilon_\theta e_z + C_1\epsilon_z^2 \tag{18}$$

$$\left. \begin{array}{l} W = A_2\epsilon_\theta^2 + B_2\epsilon_\theta e_z + C_2\epsilon_z^2 \\[2mm] \quad + D_2\epsilon_\theta^3 + E_2\epsilon_\theta^2 e_z + F_2\epsilon_\theta e_z^2 + G_2\epsilon_z^3 \end{array} \right\} \tag{19}$$

and

$$\left. \begin{array}{l} W = A_3\epsilon_\theta^2 + B_3\epsilon_\theta\epsilon_z + C_3\epsilon_z^2 + D_3\epsilon_\theta^3 \\[2mm] \quad + E_3\epsilon_\theta^2 e_z + F_3\epsilon_\theta\epsilon_z^2 + G_3\epsilon_z^3 + H_3\epsilon_\theta^4 \\[2mm] \quad + I_3\epsilon_\theta^3\epsilon_z + J_3\epsilon_\theta^2\epsilon_z^2 + k_3\epsilon_\theta\epsilon_z^3 + L_3\epsilon_z^4 \end{array} \right\} \tag{20}$$

Experiments were made *in vitro* on the thoracic aorta of 13 dogs.

The vessel was precycled and elongated by a weight while the internal

pressure was incrementally increased. One minute after each pressure

increment the dimensions were recorded. The experimental data was used

to calculate the principal stresses and strains and the constants in

equations (18), (19) and (20) evaluated.

The relationship between the three constants in equation (18)

suggested that the vessel was stiffer in the circumferential direction

than longitudinally.

The incremental moduli may be predicted from the stored energy

function and these predictions were compared with experimental

measurements. The three and twelve constant theories gave discrepancies

of up to 30% but the seven constant theory produced acceptable results.

The authors concluded that the seven constant theory was the most useful.

Fung et al (1979[45]) proposed an alternative exponential function

$$W = \frac{C'}{2} \exp\left[a_1 \varepsilon_{\theta\theta}^2 + a_2 \varepsilon_{zz}^2 + 2a_4 \varepsilon_{\theta\theta} \varepsilon_{zz}\right] \tag{21}$$

which may be rewritten as

$$W = \frac{C}{2} \exp\left[a_1 (\varepsilon_{\theta\theta}^2 - \varepsilon_{\theta\theta}^{*2}) + a_2 (\varepsilon_{zz}^2 - \varepsilon_{\theta\theta}^{*2}) + 2a_4 (\varepsilon_{\theta\theta} \varepsilon_{zz} - \varepsilon_{\theta\theta}^* \varepsilon_{zz}^*)\right] \tag{22}$$

The strains $\varepsilon_{\theta\theta}^*$ and ε_{zz}^* are the result of arbitrary applied stresses

$\sigma_{\theta\theta}^*$ and σ_{zz}^*.

The function was evaluated experimentally on arteries of rabbits.

The vessels were subjected to longitudinal extension to approximately

in vivo length and then inflated, or stretched while uninflated. The

values of $\sigma_{\theta\theta}^*$ and σ_{zz}^* were chosen to fall within the physiological range

of function of the vessel.

The overall fit between the stress-strain relations predicted from the exponential strain energy function and the experimental data was reasonably good. The fit obtained from the seven constant polynomial form discussed above was marginally better.

The major difference between the two forms of the stored energy function appeared when the constants were evaluated at different arterial sites. The polynomial constants showed wide variations and changes of sign, but the exponential constants varied less erratically.

The stored energy functions evaluated by Vaishnav et al (1972) and Fung et al (1979) are reduced forms of general three dimensional orthotropic functions, which are appropriate to the physiological type of loading used in the experimental test programmes. The loading is essentially two dimensional as the longitudinal and circumferential stresses, σ_z and σ_o, are very much greater than the radial stress σ_r. The principal stresses, and strains, are also in the direction of the axes of elastic symmetry of the vessel wall. The elimination of the radial strain, ε_r, from the three dimensional functions depend on the assumption that the vessel walls are incompressible. The walls of the large arteries are thin compared to their radius, and it is possible (Green and Adkins, 1960) to develop two dimensional stored energy functions for the two dimensional loading which occurs during function. An example of the application of such a two dimensional stored energy function, which does not depend on the assumption of incompressibility of the arterial wall can be found in Rachev (1980[46]).

4.   <u>TIME DEPENDENCE</u>

The non-linear time dependence of the arterial wall was noted by

Roy (1880[47]) who compared the behaviour with that found in rubbers.

There are numerous other reports of various aspects of time dependence

reported in the literature, and empirical descriptions of such variables

as the time course of stress relaxation.

There are two views of the non-linearity of the tissue. The first

assumes that the non-linearity of the load-extension response and time

dependence are separable leading to a single integral representation

(Fung, 1972). The alternative assumption that this simplification is

unjustified leads to the use of formulations for non-linear viscoelasticity

of the Green-Rivlin type. The former assumption will be discussed in this

section and multiple integral representations in Section 5.

4.1.   Single Integral Representations

The stress $\sigma(t)$ at time t after the application of a step extension

$\lambda(o)$ to a viscoelastic solid is given by

$$\sigma(t) = \lambda(o) K(\lambda,t) \tag{23}$$

where $K(\lambda,t)$ is the relaxation function.

If the relaxation function is separable in extension and time this

reduces to

$$K(\lambda,t) = G(t) K^{(e)}(\lambda) \tag{24}$$

The nomenclature follows Fung (1972) and $G(t)$ is a reduced

relaxation modulus and $K^{(e)}(\lambda)$ an elastic response which is a function only of the extension ratio λ and is called the elastic response. If it is assumed that a suitable modified superposition principle applies then the stress produced by a continuous stretch history is analogous to the Linear Boltzman's superposition integral (see Flügge, 1975[48]) and

$$\sigma(t) = \int_{-\infty}^{t} G(t-\tau) \frac{\partial T^{(e)} |\lambda(\tau)|}{\partial \lambda} \frac{\partial \lambda(\tau)}{\partial \tau} d\tau$$

$$\text{or} \quad \sigma(t) = \int_{-\infty}^{t} G(t-\tau) \dot{T}^{(e)}(\tau) d\tau \tag{25}$$

Equation 25 may be inverted to obtain the creep function relating the time dependent extension and the applied step stress. In addition the use of equation 25 leads to the prediction of the response to the application of a sinusoidally varying deformation.

The elastic response $T^{(e)}(\lambda)$ can be identified with the characterisations of the overall load-extension response discussed in Section 3.

4.2   Descriptions of Time Dependence

The time dependence of the tissues have been described in terms of both discrete time constraints and continuous relaxation spectra.

Discrete time constraints lead to the characterisation of the stress relaxation behaviour in terms of the sums of time varying exponentials. That is

$$G(t) = C_1 \exp(-t/\tau_1) + C_2 \exp(-t/\tau_2) + \ldots$$

$$\text{or} \quad G(t) = \sum_i C_i \exp(-t/\tau_i) \tag{26}$$

Fung (1972) used a normalised relaxation modulus and equation 5 becomes

$$G(t) = \sum_i C_i \exp(-t/\tau_i)/\sum_i C_i \tag{27}$$

There are three major drawbacks to using such a discrete time constant characterisation.

It is usually impossible to obtain unique values of the parameters, and the same data can give rise to quite different exponential terms if the method of analysis is changed. Lanczos (1956[49]) discussed this point at some length, with examples, and concluded that the difficulty lay in the extreme sensitivity of the parameters to small changes of the data. He thought that the use of least squares or other statistics would not provide a remedy.

The second drawback is the tendency for those deriving the time constraints to return to the spring and dashpot mechanical analogues which were used as conceptual aids by Kelvin and Maxwell and to identify these imaginary mechanical analogues with real structures in the tissue (e.g. Wiedehielm, 1965).

The greatest disadvantage is that models incorporating discrete time constants generally predict a marked frequency dependence of the dynamic viscoelastic parameters. This can be demonstrated by the behaviour of the simplest model showing the essential features of the arterial wall. The tissue shows an instantaneous elastic response and an equilibrium stress after the application of a step deformation, and the simplest model showing these features is the three parameter (or

standard) viscoelastic solid. The relaxation modulus of such a material
is given by

$$G(t) = G_I + G_e|1 - \exp(-t/\tau)|$$ (28)

where G_I and G_e are the instantaneous and equilibrium moduli. Using the
general relations between the viscoelastic parameters it is possible to
use the relationship in equation 28 to predict the storage and loss
moduli $G'(w)$ and $G''(w)$ which the material will display in response to a
sinusoidal deformation. These are given by:

$$G'(w) = G_e + (G_I - G_e) \frac{w^2\tau^2}{1 + w^2\tau^2}$$

$$G''(w) = (G_I - G_e)w\tau/1 + w^2\tau^2$$ (29)

Both parameters are frequency dependent, with the phase angle
showing a peak value at an angular frequency w equal to $1/\tau$. This
behaviour is at variance with the frequency insensitivity of the phase
angle which has been shown experimentally (Section 2).

Continuous relaxation spectra replace the discrete time constants
with a continuous function. For such a system the stress relaxation
modulus $G(t)$ in equation 26 becomes

$$G(t) = G_e + \int_0^\infty A(\tau) \exp(-t/\tau) \, d\tau$$ (30)

where $A(\tau)$ is the continuous relaxation spectrum. Alternatively the
logarithmic spectrum $H(\tau)$ may be used (Ferry 1970[50])

$$G(t) = G_e + \int_{-\infty}^{+\infty} H(\tau) \exp(-t/\tau) d\ln\tau$$ (31)

The value of $H(\tau)$, as a function of time, may be obtained from the slope of a plot of the stress relaxation modulus against the logarithm of elapsed time. To a first approximation this is (Alfrey, 1948[51])

$$H(\tau) = - \frac{1}{2.303} \left| \frac{dG(t)}{d\ell n t} \right|_{t=\tau} \qquad (32)$$

In terms of stress $\sigma(t)$ and initial extension $\lambda(o)$ equation 12 is

$$H(\tau) = - \frac{1}{2.303 \, \lambda(o)} \left| \frac{d\sigma(t)}{d\ell n t} \right|_{t=\tau} \qquad (33)$$

The relaxation in arteries is such that the stress falls linearly with the log of time for several decades. This means both that the approximation implicit in equations 31 and 32 is a good one, and that the logarithmic spectrum is particularly simple, consisting of a spectrum of constant value over the time region for which the relaxation is linear and zero outside this. This spectrum is usually known as a box spectrum, and its properties have been investigated both for polymers and for tissues (Fung, 1972; Barbenel et al, 1973[52]).

The spectrum can be characterised as having the value

$$\left. \begin{array}{ll} H(\tau) = H(o) & \tau_1 < t < \tau_2 \\[2mm] \quad\;\; = 0 & t < \tau_1; \; t > \tau_2 \end{array} \right\} \qquad (34)$$

Substituting equation 14 into equation 11 yields

$$G(t) = G_e + H(o) \int_{\tau_1}^{\tau_2} \exp(-t/\tau) \, d\ell n \tau \qquad (35)$$

which can be integrated to yield (Tobolsky, 1960[53])

$$G(t) = G_e + H(o) \left| E_i(-t/\tau_1) - E_i(-t/\tau_2) \right| \qquad (36)$$

where E_i is the exponential integral.

The series expansion for E_i is

$$E_i(x) = \gamma + \ln x + \left(\frac{x}{1.1!} - \frac{x^2}{2.2!} + \ldots\right) \tag{37}$$

where γ is Euler's constant.

Thus for $t < \tau_1 \ll \tau_2$

$$G(t) = G_e + H(o) \ln (\tau_2/\tau_1) \tag{38}$$

for $\tau_2 < t$

$$G(t) = G_e \tag{39}$$

and in the middle range where $\tau_1 < t < \tau_2$

$$G(t) = G_e - H(o) (\ln 1/\tau_2 - \ln t) \tag{40}$$

which reproduces the experimental stress relaxation behaviour.

The upper limit τ_2 of the spectrum can be obtained from the time, t_2, at which the extrapolated linear central portion of the relaxation curve is equal to the equilibrium modulus,

$$\begin{aligned} \tau_2 &= t_2 \exp \gamma \\ &= 1.781 \ t_2 \end{aligned} \tag{41}$$

and similarly

$$\tau_1 = 1.781 \ t_1 \tag{42}$$

These relationships can be used to obtain the box parameters from the experimental stress relaxation curve and to predict the dynamic storage and loss moduli from the relationships (Olofsson, 1974[54])

$$G'(w) = G_e + \int_{-\infty}^{+\infty} \left| H(o)w^2\tau^2 \Big/ (1+w^2\tau^2) \right| d\ell n \ \tau$$

$$= G_e + \frac{H(o)}{2} \ \ell n \ \frac{1+w^2\tau_2^2}{1+w^2\tau_1^2} \tag{43}$$

and

$$G''(w) = \int_{-\infty}^{+\infty} \left| H(o)\tau \Big/ (1+w^2\tau^2) \right| d\ell n \ \tau$$

$$= H(o) \ (\tan^{-1} \ w\tau_2 - \tan^{-1} \ w\tau_1) \tag{44}$$

In the time region of the non zero portion of the spectrum the phase angle, δ, is given by

$$\delta = \tan^{-1} \left| \frac{\pi}{2} - \frac{1}{G_e/H(o), \ + \ell n(w\tau_2)} \right|$$

$$= \tan^{-1} \ \frac{\pi \ H(o)}{2G_e} \tag{45}$$

The general form of the dynamic response predicted by the box spectrum is similar to that found experimentally, and this led to the suggestion of its use for describing time dependence in soft connective tissues.

Tanaka and Fung (1974) showed that it is applicable to the aorta and that the values of τ_2 varied from 94 s to 2480 s, with a major site dependence. The values of τ_1 were all less than 1 s. These values also imply that the load-extension response would be independent of strain rate in the range $10^{-3} - 1 \ s^{-1}$.

The single integral relationship between stress, and strain and time produced by both the discrete time constant and continuous spectrum model will predict a progressive alteration in response on repetitive straining. No detailed analysis of the magnitude of this predicted

preconditioning has been made, but the nature of the experimental stress
relaxation, which has a prolonged course makes it probable that the
predicted preconditioning will be small, unlike the major changes found
experimentally.

In order to incorporate either model of time dependence into the
single integral representation of equation 25, it is essential to
confirm that the viscoelastic parameters are independent of the strain
magnitude. It is not clear that a critical test of this very important
condition has been made for artery, although it appears not to be
satisfied for *in vivo* skin.

These two unresolved features of the single integral representation
has led to the investigation of multiple integral forms.

4.3 Multiple Integral Representations

Green and Rivlin (1957[55], 1960[56]) proposed a multiple integral
representation for the response of a general non-linear viscoelastic
material. For one dimensional behaviour this reduces to (Lockett, 1972[57])

$$
\left.
\begin{aligned}
\sigma(t) &= \int_{-\infty}^{t} G_1(t-\tau_1)\dot{\epsilon}(\tau_1)d\tau_1 + \iint_{-\infty}^{t} G_2(t-\tau_1,t-\tau_2)\dot{\epsilon}(\tau_1)\dot{\epsilon}(\tau_2)d\tau_1\,d\tau_2 \\
&+ \iiint_{-\infty}^{t} G_3(t-\tau_1,t-\tau_2,t-\tau_3)\dot{\epsilon}(\tau_1)\dot{\epsilon}(\tau_2)\dot{\epsilon}(\tau_3)d\tau_1\,d\tau_2\,d\tau_3 + \ldots
\end{aligned}
\right\} (46)
$$

where G_1, G_2 and G_3 ... are first, second, third ... order kernels; it
can be assumed without any loss of generality that the kernels are
symmetric with respect to their arguments. The term ϵ is a suitable
strain measure.

To be useful the series must be finite and is usually terminated after the third term.

The response of such a material can best be analysed by utilising the step function Δt, which has the properties

$$\Delta(t) = 1 ; \qquad t > 0$$

$$\Delta(t) = 0 ; \qquad t < 0$$

The application of a step strain $\varepsilon(0)\Delta t$ produces a stress $\sigma(t)$ given by:

$$\sigma(t) = \varepsilon(0)G_1(t) + \varepsilon^2(0)G_2(t,t) + \varepsilon^3(0)G_3(t,t,t) \qquad (47)$$

It is apparent that the material is non-linear.

The response to a single one step test is given in the above equation and it is clear that the response to a single step test is not sufficient to determine the material constraints. The test may be repeated with two additional strain values $\varepsilon(1)\Delta t$ and $\varepsilon(2)\Delta t$ and from the three test results it is possible to calculate the values of

$$G_1(t), \; G_2(t,t) \text{ and } G_3(t,t,t).$$

In order to obtain G_2 and G_3 when their arguments are unequal, it is necessary to perform multiple step tests.

The second order kernel G_2 can be evaluated using two step tests $\varepsilon(t) = \varepsilon(3)\Delta t + \varepsilon(4)\Delta(t-t_1)$. The response is:

$$
\begin{aligned}
\sigma(t) = \; & \varepsilon(3)G_1(t) + \varepsilon^2(3)G_2(t,t) + \varepsilon^3(3)G_3(t,t,t) \\
& + \varepsilon(4)G_1(t-t_1) + \varepsilon^2(4)G_2(t-t_1,\; t-t_1) \\
& + 2\varepsilon(3)\varepsilon(4)G_2(t,\; t-t_1) + \varepsilon^3(4)G_3(t-t_1,\; t-t_1,\; t-t_1) \\
& + 3\varepsilon(3)\varepsilon(4)\left| \varepsilon(3)G_3(t,t,t-t_1) + \varepsilon(4)G_3(t,\; t-t_1,\; t-t_1)\right|
\end{aligned}
\qquad (48)
$$

The first six terms are known from the single step tests and can be evaluated. In order to evaluate the last three terms it is necessary to carry out two additional two step tests in which t_1 is kept constant. This allows the evaluation of values of three kernels

i.e. $G_2(t, t-t_1)$; $G_3(t, t, t-t_1)$ and $G_3(t, t-t_1, t-t_1)$.

Thus G_2 has been determined for unequal arguments, but only for one value of the difference in argument. The number of values of the kernel $G_2(t, t-k)$ required to determine the complete function between k=0 and the value of k at which an equilibrium value occurs depends on the form of $G_2(t, t-k)$. This can only be determined by experiment, and Lockett has suggested that a minimum of ten values of k is required.

As a result of the experiments to determine $G_2(t, t-k)$ the values of third order kernel are known when two of the arguments are equal. In order to determine the value when all three are unequal, $G_3(t, t-k, t-\ell)$ it is necessary to carry out a single three step test. Once again it is necessary to carry out multiple three step tests in order to cover a range of values for k and ℓ. Lockett (1972) has calculated the number of tests required to evaluate all three constants at n discrete values of k and ℓ . For n = 10, which Lockett suggests is a reasonable number, the test programme requires 78 tests, and ideally this should be increased to obtain replicate results.

The one dimensional relationship in equation 46 can be generalised to produce a three dimensional theory for anisotropic materials. Cheung and Hsaio (1972[58]) presented a theoretical and experimental analysis based on the assumption that the artery wall was transversely isotropic, for which there would appear to be little justification. A similar

study was made by Young et al (1977[59]) who assumed the wall to be

incompressible and to be orthotropic.

The number of kernel functions depends on the order of the integral

equation. Young et al investigated a second order theory, and showed

for this case orthotropy required the evaluation of ten kernel functions.

By dropping second order kernel functions, a simpler first order equation

containing only four kernels was obtained. In the latter case it

appears the non-linearity of the response is determined by the strain

measure chosen. The lack of second order terms, and therefore the

presence of single integrals only, limits the nature of preconditioning

the theory can describe.

Tests were made on excised segments of the descending aorta of dogs,

and before testing the vessel was returned to its *in vivo* length and

repeatedly pressurised to a pressure of 19 kPa (200 cm H_2O), which

suggests that the vessel was, in fact, preconditioned before testing

started. To evaluate the kernel functions data was obtained on stress

relaxation at 38 states of strain, which were obtained by longitudinal

extension and inflation of the vessel segment. Relaxation was measured

for 100s. The results suggest that each step was considered in

isolation with results being presented in terms of a single time

argument.

The experimental data from which the parameters were obtained were,

not unsurprisingly, well described by both theories. The critical test

of their applicability is the prediction and experimental verification of the response to a different state of strain.

4.4   Time Dependence in Biological Tissues

The time dependence of biological tissues has been described in
terms of classical viscoelastic theory. The appearance of terms of the
nature of $\phi(t-\tau)$ leads to a hereditary behaviour, such that the
influence of events occurring a long time before the current time at
which measurements are made will gradually diminish. It appears from
the response of preconditioned tissues that the application of a large
strain to the specimen may influence the current state of stress as much
as the current strain. It is not clear that such behaviour can be
contained within classical viscoelasticity. An alternative approach was
used by Chu and Blatz (1972[60]), who postulated a cumulative microdamage
process to describe the mechanical behaviour of mesentery. The
irreversibility of the microdamage limits the applicability of the theory,
but investigations of alternatives to viscoelastic time dependence may
suggest how the effects of preconditioning may be incorporated into a
general theory of time dependence for tissues.

5. REFERENCES

1. McDonald, D.A., Blood flow in arteries, Edward Arnold Ltd.,
    London, 1974.

2. Caro, C.G., Pedley, T.J., Schroter, R.C. and Seed, W.A., The
    Mechanics of the circulation. Oxford University Press,
    Oxford, 1978.

3. Iberall, A.S., Anatomy and steady flow characteristics of the
    arterial system, *Math. Bioscience* 1, 375-395, 1967.

4. d'Arcy, W. Thompson, On growth and form, MacMillan, London,
    pp. 948-957, 1945.

5. Bloom, W. and Fawcett, D.W., A textbook of histology. W.B. Saunders,
    Philadelphia, 1975.

6. Benninghoff, A., Blutgafasse and Herz in 'Handbuch der
    Microkopischen Anatomie', Springer Verlag, Berlin, Vol. Vl/1,
    p.1-225, 1930.

7. Wolinsky, H. and Glagov, S., A lamellar unit of aortic medial
    structure and function in mammals. *Circulation Res.* 20,
    99-111, 1967.

8. Burton, A.C., Physical principles of circulatory phenomena.
    Handbook of physiology, Amer. Physiol. Soc., Washington,
    Vol 1/2, pp. 85-106, 1962.

9. Folkow, B. and Neil, E., Circulation, Oxford University Press,
    London, 1971.

10. Learoyd, B.M. and Taylor, M.G., Alteration with age in the
     viscoelastic properties of human arterial wall. *Circulation Res.*
     18, 278-291, 1966.

11. Fung, Y.C., Biomechanics, Mechanical properties of living tissues,
     Springer Verlag, New York, 1981.

12. Tanaka, T.T. and Fung, Y.C., Elastic and inelastic properties of
     the canine aorta and their variation along the aortic tree.
     J. Biomech. 7, 357-370, 1974.

13. Collins, R. and Hu, W.C., Dynamic deformation experiments on aortic
     tissue, *J. Biomech.* 5, 333-337, 1972.

14. Wiederhielm, C.A., Distensibility characteristics of small blood
     vessels. *Fed. Proc.* 24, 1075-1084, 1965.

15. Patel, D.J. and Fry, D.L., The elastic symmetry of arterial
     segments in dogs. *Circ. Res.* 24, 1-8, 1969.

16. Patel, D.J. and Vaishnav, R.N., Basic hemodynamics and its role in
     disease processes, University Park Press, Baltimore, 1980.

17. Lawton, R.W., The thermoelastic behaviour of isolated aortic
     strips of the dog. *Circ. Res.* 3 403-408, 1954.

18. Carew, T.E., Vaishnav, R.N. and Patel, D.J., Compressibility of the
     arterial wall. *Circulation Res.* 22, 61-68, 1968.

19. Bergel, D.H., The properties of blood vessels in 'Biomechanics -
     it's foundations and objectives' ed. Fung, Y.C., Persone N.
     and Anliker M. Prentice-Hall, New Jersey, pp. 105-139, 1972.

20. Veronda, D.R. and Westmann, R.A., Mechanical characterisation of
     skin - finite deformations. *J. Biomech.* 3, 111-124, 1970.

21. Stark, H.L. and Al-Haboubi, A., The relationship of width, thickness,
     volume and load to extension for human skin, *In Vitro,
     Engineering in medicine* 9, 179-183, 1980.

22. Tickner, E.G. and Sacks, A.H., A theory for the static elastic
     behaviour of blood vessels. *Biorheology* 4, 151-168, 1967.

23. Wolinsky, H. and Glagov, S., Structural basis for the static
     mechanical properties of the aortic media. *Circulation Res.* 14,
     400-413, 1964.

24. Carton, R.W., Dainauskas, J. and Clark, J.W., Elastic properties
     of single elastic fibres. *J. Appl. Physiol.* 17, 547-551, 1962.

25. Jenkins, R.B. and Little, R.W., A constitutive equation for
     parallel fibred elastic tissue. *J. Biomech.* 7, 397-402, 1974.

26. Gosline, J.N., The physical properties of elastic tissue in
     'International review of connective tissue research'. ed.
     Hall D.A. and Jackson, D.S. Academic Press, London Vol. 7.
     pp. 211-250, 1976.

27. Haut, R.C. and Little, R.W., A constitutive equation for collagen
     fibres. *J. Biomech.* 5, 289-298, 1972.

28. Dobrin, P.B. and Rovick, A.A., Influence of vascular smooth muscle
     on contractive mechanics and elasticity of arteries. *Amer. J.
     Physiol.* 217, 1644-1651, 1969.

29. Harkness, M.L.R., Harkness, R.D. and MacDonald, D.A., The collagen
     and elastin content of the arterial wall in dogs. *Proc. Roy.*
     Soc. B. <u>146</u>, 541-551, 1957.

30. McCrum, N.G. and Dorrington, K.L., The bulk modulus of solvated
     elastin. *J. Mat. Sci.* <u>11</u>, 1367-1368, 1976.

31. Roach, M.R. and Burton, A.C., The reason for the shape of the
     distensibility curves of arteries, *Can. J. Biochem. Phsiol.* <u>35</u>,
     681-690, 1957.

32. Hoffman, A.S., Grande, L.A., Gibson, P., Park, J.B., Daly, C.H.,
     Borstein, P. and Ross, R., Preliminary studies on mechanochemical
     -structure relationships in connective tissues using enzymolysis
     techniques, in "Perspectives in biomedical engineering", ed.
     Kenedi, R.M., MacMillan, London, pp.173-176, 1973.

33. Patel, D.J., Janicki, J.S. and Carew, T.E., Static anisotropic
     elastic properties of the aorta in living dogs. *Circulation Res.*
     <u>25</u>, 765-779, 1969.

34. Fung, Y.C., Stress-Strain history relations of soft tissue in
     simple elongation in 'Biomechanics - its foundations and
     objectives' ed. Fung, Y.C. Perrone, N. and Anliker, M.
     Prentice-Hall, New Jersey, pp. 181-208, 1972.

35. Vito, R., A note on arterial elasticity. *J. Biomechs.* <u>1</u>, 3-12, 1973.

36. Biot, M.A., Mechanics of incremental deformation. John Wiley,
     New York, 1965.

37. Bergel, D.H., The static elastic properties of the arterial wall.

      J. Physiol. 156, 458-469, 1961.

38. Patel, D.J. and Fry, D.L., Longitudinal tethering of arteries in

      dogs. Circ. Res. 19, 1011-1021, 1966.

39. Fung, Y.C., Rheology of blood vessels in 'microcirculation'

      ed. Kaley, G. and Altura, B.M. University Park Press,

      Baltimore, Vol. 1 pp. 299-324, 1977.

40. Hardung, V., Die bedentung der anisotropie and inhomogenitat bei

      der bestummung der elastizitat der blutgefasse II.

      Angiologica 1, 185-196, 1964.

41. Bergel, D.H. and Schultz, D.L., Arterial elasticity and fluid

      dynamics in 'Progress in biophysics and molecular biology'.

      ed. Butler, J.A.V. and Noble, D. Pergamon Press, Oxford 22,

      3-36, 1971.

42. Green, A.E. and Adkins, J.E., Large elastic deformations and

      non-linear continuum mechanics, Clarendon Press, Oxford, 1960.

43. Green, A.E. and Zerna, W., Theoretical Elasticity. Clarendon Press,

      Oxford, 1954.

44. Vaishnav, R.N., Young, J.T., Janicki, J.S. and Patel, D.J.,

      Non linear anisotropic elastic properties of the canine aorta.

      Biophys. J. 12, 1008-1027, 1972.

45. Fung, Y.C., Fronek, K and Patitucci, P., Pseudo elasticity of

      arteries and the choice of its mathematical expression.

      Am. J. Physiol. 237, H620-H631, 1979.

46. Rachev, A.I., Effects of transmural pressure and muscular activity
     on pulse waves in arteries. *ASME Journal of Biomechanical
     Engineering* 102, 119-123, 1980.

47. Roy, C.S., The elastic properties of the arterial wall, *J. Physiol.*
     3, 125-159, 1880.

48. Flügge, W., Viscoelasticity. 2nd Ed. Springer-Verlag, Berlin, 1975.

49. Lanczos, C., Applied analysis. Prentice-Hall, New Jersey, 1956.

50. Ferry, J.D., Viscoelastic properties of polymers. John Wiley,
     New York, 1970.

51. Alfrey, T., Mechanical behaviour of high polymers. Interscience,
     New York, 1948.

52. Barbenel, J.C., Evans, J.H. and Finlay, J.B., Stress-strain-time
     relations for soft connective tissues. 'Perspectives in
     biomedical engineering'. ed. Kenedi, R.M., Macmillan, London,
     pp. 165-172, 1973.

53. Tobolsky, A.V., Properties and structure of polymers, John Wiley,
     New York, 1960.

54. Olofsson, B., Comparison of stress-activated models and linear
     spectral models for visco-elasticity. Rheol. Acta. 13, 78-85,
     1974.

55. Green, A.E. and Rivlin, R.S., The mechanics of non-linear materials
     with memory. *Arch. Rat. Mech. Anal.* 1, 1-21, 1957.

56. Green, A.E. and Rivlin, R.S., The mechanics of non-linear materials
     with memory. *Arch. Rat. Mech. Anal.* 4, 387-404, 1960.

57. Lockett, F.J., Non-linear viscoelastic solids. Academic Press,
     London, 1972.

58. Cheung, J.B. and Hsiao, C.C., Non-linear anisotropic viscoelastic
     stresses in blood vessels. *J. Biomechics*. 5, 607-619, 1972.

59. Young, J.T., Vaishnav, R.N. and Patel D.J., Non-linear anisotropic
     viscoelastic properties of canine arterial segments.
     J. Biomechs. 10, 549-559, 1977.

60. Chu, B.M. and Blatz, P.J., Cumulative microdamage models to
     describe the hysteresis of living tissue. *Ann. Biomed. Engng*. 1,
     204-211, 1972.

CHAPTER III
DYNAMICS OF FLUID FILLED TUBES

J.B. Haddow
Department of Mechanical Engineering
T.B. Moodie, R.J. Tait
Department of Mathematics
University of Alberta
Edmonton, Alberta, Canada

1. INTRODUCTION

1.1 Introduction

Models for describing pressure pulse propagation in arteries can,
with few exceptions, be separated into two categories depending upon
the emphasis placed on the fluid mechanics or the solid mechanics.
Models in the first category employ ideas first put forward by Euler
(1844) who proposed combining the equation of motion for an inviscid
fluid together with the continuity equation and a third equation relating the pressure in the tube to its cross-sectional area. Euler's
equations were first solved in the context of blood flow by Lambert
(1958) who used a pressure-area relation based on experimental data
rather than the one proposed by Euler. These ideas have been considerably refined and developed to employ general pressure-area relations,

include outflow through porous walls (Rudinger, 1966), and study the

development of shocks (Rudinger, 1970), (Teipel, 1973), (Forbes, 1979).

The second category, and the one we are concerned with is comprised

of those models employing a version of the shell equations for the tube

wall together with equations of motion for the fluid. An excellent re-

view of these models up to 1966 is included in the survey article by

Skalak (1966). Significant contributions in this category have been

made by Lamb (1898), Witzig (1914), Klip (1962), Rubinow and Keller (1968),

(1971), (1978), and many others. All of the models in this category are

analyzed in essentially the same way although the amount of detailed in-

formation obtained varies greatly from author to author. Solutions are

assumed in the form of travelling periodic waves and frequency equations

are obtained and plotted together with plots for the various mode shapes.

In no instance are either initial or boundary value problems solved for

models in this category.

It is our aim to outline a method of solution for an initial value

problem for fluid-filled tube models in this second category. We do not

make the claim that our results in their present form are directly ap-

plicable to the propagation of pressure pulses in arteries but rather

that these methods do indicate a viable means of analyzing the transient

response of various models in this category. The information so ob-

tained can then be compared with experimental results to test the

ability of such models to explain observed phenomena.

The first model considered is a straight, uniform, thin walled,

tethered, cylindrical, elastic tube filled with an incompressible inviscid fluid. Viscosity terms can be included in the analysis with additional complication, however, preliminary work has shown that for small disturbances superimposed on the flow in elastomer tubes the damping due to viscous effects in the tube wall is large compared with that due to fluid viscosity when parameters appropriate for biological applications are considered. We first present the governing equations of the problem taking account of shear deformation of the tube wall and discuss the validity of neglecting the shear deformation. Later we consider the governing equations with shear neglected and extend the analysis to consider viscoelastic tubes.

Although the theory is given for a linearly elastic or viscoelastic tube, with small modification it can be adapted to consider small perturbations of a tube subjected to finite deformation.

Results are presented for a tube with wall thickness to mean radius ratio of 0.05 although the theory is valid for ratios up to approximately 0.15. As this ratio is increased the effect of shear deformation of the tube wall becomes more important.

A slightly different approach to the problems considered here has been presented by Tait, Moodie and Haddow (1981).

1.2 Formulation of Problem

A uniform thin-walled tube with its axis horizontal and in the x-direction is considered. The radius, density, and wall thickness of the tube are R, γ, and h, respectively, and ρ is the density of the fluid. The radius R is taken to be the inside radius of the tube with only a small error introduced by using R in the shell equations in place of the mean radius R + h/2. The distinction between the mean radius and inside radius of the tube could be incorporated in the analysis with some extra complication but the effect on the results was found to be negligible for h/R < 0.1.

It is assumed that an axially symmetric perturbation of the system results in axial and radial perturbation displacements $u_x(r,x,t)$ and $u_r(r,x,t)$, respectively, of the fluid. Corresponding velocity perturbations are $\partial u_x/\partial t$ and $\partial u_r/\partial t$. It is further assumed that the pressure on the outside of the tube is uniform and that the pressure perturbation is small enough that a linear theory is valid. Since a tethered tube is considered, there is no axial displacement of the tube wall. It is well known that blood vessels in situ operate under considerable axial constraint. We include shear deformation in the analysis and incorporate it along with rotatory inertia into a simple shell theory for the fluid-filled tube.

In order to illustrate the method of solution of initial value problems we will consider an unbounded tube to be perturbed by an initial distribution of external pressure $p_e(x)$ which is suddenly removed at time t = 0. For definiteness we consider the external loading

$$p_e = q\delta(x) \, [1 - H(t)] \qquad\qquad (1.2.1)$$

where H(t) is the unit step function and $\delta(x)$ is the Delta function. This is a concentrated circumferential loading of intensity q per unit length of circumference as shown in Fig. 1.2.1.

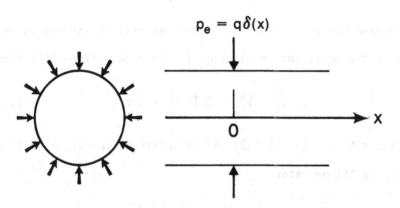

$$p_e = q\delta(x)$$

Fig. 1.2.1 Concentrated circumferential loading
of intensity q per unit length of circumference

1.3 Constitutive Equations of Tube Material

The tube wall is assumed to be isotropic, incompressible and linearly elastic or viscoelastic. Consequently, for the elastic material Poisson's ratio $\nu = 0.5$ and the constitutive equation is

$$\underset{\sim}{s} = 2\mu\underset{\sim}{e} \, , \qquad\qquad (1.3.1)$$

where $\underset{\sim}{\sigma}$ and $\underset{\sim}{e}$ are the stress deviator and strain tensors, respectively, and μ is the shear modulus. The assumption of incompressibility is appropriate for tubes in biological systems.

The viscoelastic model considered is the so-called standard viscoelastic material. A spring-damper analogue model for this material is shown in Fig. 1.3.1. In Fig. 1.3.1 the stiffness μ is analogous to the impact or unrelaxed modulus of rigidity, η represents the viscosity, and $0 < m < 1$. The relaxation time τ is given by

$$\tau = \frac{\eta}{\mu(1-m)} \, , \tag{1.3.2}$$

and the relaxed modulus by $m\mu$. Since the material is assumed to be in-compressible the constitutive equation is given in differential form by

$$(\tau \frac{\partial}{\partial t} + 1)\underset{\sim}{s} = 2\mu(\tau \frac{\partial}{\partial t} + m)\underset{\sim}{e} \, , \tag{1.3.3}$$

where t is the time. Eq. (1.3.3) can be obtained from eq. (1.3.1) by replacing μ by the operator

$$\mu(\tau \frac{\partial}{\partial t} + m)/(\tau \frac{\partial}{\partial t} + 1) \, . \tag{1.3.4}$$

The integral form of the constitutive eq. (1.3.3) is

$$\underset{\sim}{s} = \int_{-\infty}^{t} G(t-t') \frac{d\underset{\sim}{e}(t')}{dt'} \, dt' \, ,$$

where the relaxation modulus G is given by

$$G(t) = 2\{m\mu + (1-m)\mu \, \exp(-t/\tau)\} \, .$$

As $m \to 1$, eq. (1.3.3) approaches the constitutive equation (1.3.1) for a Hookean solid and as $m \to 0$, that for a Maxwell fluid.

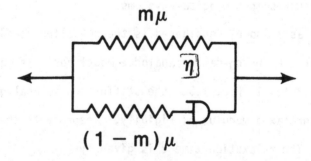

Fig. 1.3.1 Spring-damper analogue for
standard viscoelastic material

2. GOVERNING EQUATIONS

2.1 Tube Equations

When shear deformation and rotatory inertia of the tube wall are taken into account, the equations of motion of the tube wall for a tethered elastic tube, as obtained from a simplified shell theory, are:

$$D \frac{\partial^2 \psi}{\partial x^2} + \mu Q h (\frac{\partial w}{\partial x} - \psi) - \frac{\gamma h^3}{12} \frac{\partial^2 \psi}{\partial t^2} = 0 , \qquad (2.1.1)$$

$$kw + \mu Q h (\frac{\partial \psi}{\partial x} - \frac{\partial^2 w}{\partial x^2}) + \gamma h \frac{\partial^2 w}{\partial t^2} - P_w = 0 , \qquad (2.1.2)$$

where γ is tube wall density, w is the radial displacement perturbation of the tube wall, ψ is the part, due to bending, of the slope $\partial w / \partial x$ of the tube wall, $p_w = p_i - p_e$, p_i and p_e are the internal and external pressures at the wall, respectively, Q is the shear deflection coefficient, and

$$K = \frac{Eh}{R^2(1-\nu^2)} , \quad D = \frac{Eh^3}{12(1-\nu^2)} ,$$

where E is Young's modulus. Except for notation, eqs. (2.1.1) and (2.1.2) are the same as those used by Forrestal and Herrmann (1965). Since it is assumed that the tube wall is incompressible, Poisson's ratio $\nu = 0.5$, and $E = 3\mu$. The shear coefficient Q is taken as Q = 0.91 which is the value suggested by Herrmann and Mirsky (1956) for $\nu = 0.5$.

If shear deformation of the tube wall is neglected the single equation of motion of the tube wall for an elastic tube is

$$D \frac{\partial^4 w}{\partial x^4} + kw = p_w - \gamma h \frac{\partial^2 w}{\partial t^2} + \frac{\gamma h^3}{12} \frac{\partial^4 w}{\partial x^2 \partial t^2} . \qquad (2.1.3)$$

The last term on the right hand side of eq. (2.1.3) represents the rotatory inertia. Actually, the effect of rotatory inertia is less than that of shear deformation, however, it can be incorporated without involving an additional equation of motion.

When the tube wall is composed of incompressible standard material, the shear modulus μ in eqs. (2.1.1), (2.1.2) and (2.1.3) is replaced by the differential operator (1.3.4).

2.2 Fluid Equations

The effect of fluid viscosity is neglected since it is assumed that damping arises mainly from the viscoelastic properties of the tube wall. Consequently the equations of motion of the fluid are the linearized Euler equations,

$$\rho \frac{\partial^2 u_x}{\partial t^2} = - \frac{\partial p}{\partial x} , \qquad (2.2.1)$$

$$\rho \frac{\partial^2 u_r}{\partial t^2} = - \frac{\partial p}{\partial r} , \qquad (2.2.2)$$

where r is the radial coordinate, ρ is the fluid density, $p = p(r,x,t)$ is the pressure perturbation. The internal pressure at the wall sur-face is given by $p_i(x,t) = p(R,x,t)$. In addition to eqs. (2.2.1) and (2.2.2) the continuity equation

$$\frac{1}{r}\frac{\partial}{\partial r}(ru_r) + \frac{\partial u_x}{\partial x} = 0 \ , \tag{2.2.3}$$

is required. Eliminating p from the Euler equations (2.2.1) and (2.2.2) gives

$$\frac{\partial^4 u_x}{\partial t^2 \partial x \partial r} = \frac{\partial^4 u_r}{\partial t^2 \partial x^2} \tag{2.2.4}$$

and then eliminating u_x from eq. (2.2.4) and the continuity equation (2.2.3) gives

$$\frac{\partial^4 u_r}{\partial r^2 \partial t^2} - \frac{1}{r^2}\frac{\partial^2 u_r}{\partial t^2} + \frac{1}{r}\frac{\partial^3 u_r}{\partial r \partial t^2} + \frac{\partial^4 u_r}{\partial t^2 \partial x^2} = 0 \ . \tag{2.2.5}$$

Eliminating u_r and u_x from eqs. (2.2.1), (2.2.2) and (2.2.3) gives

$$\frac{\partial^2 p}{\partial r^2} + \frac{1}{r}\frac{\partial p}{\partial r} + \frac{\partial^2 p}{\partial x^2} = 0 \ . \tag{2.2.6}$$

2.3 Nondimensionalization Scheme

Before solving the governing equations it is convenient to introduce the following nondimensional variables:

$$\bar{r} = r/R \ , \ \bar{x} = x/R \ , \ \bar{t} = tc/R \ ,$$

$$(\bar{u}_r, \bar{u}_x, w) = (u_r, u_x, w)/R \ , \ \alpha = h/r \ , \tag{2.3.1}$$

$$(\bar{p}, \bar{p}_1, \bar{p}_e) = (p, p_1, p_e)/\rho c^2 \ ,$$

where c is the classical Korteweg-Moens wave speed and is given by

$$c = \left(\frac{Eh}{2\ R(1-\nu^2)} \right)^{1/2} \ ,$$

and for $\nu = 0.5$, which is the case considered here

$$c = \left(\frac{2\mu\alpha}{\rho} \right)^{1/2} \ . \qquad\qquad (2.3.2)$$

Often when dynamic viscoelastic problems are considered the nondimensionalization of quantities which have the dimension of time is based on the relaxation time τ, however, in this work the nondimensionalization is based on the wave speed (2.3.2) and the radius R so that the nondimensional relaxation time is

$$\bar{\tau} = \tau c/R \ . \qquad\qquad (2.3.3)$$

In eq. (1.2.1) the nondimensional forms of q and the Delta function are

$$\bar{q} = \frac{q}{\rho c^2 R} \ , \ \delta(\bar{x}) = \delta(x)\ R \ , \qquad\qquad (2.3.4)$$

respectively.

Two additional quantities arise later, ω, the circular frequency, and k, the wave number, and the nondimensional forms are

$$\bar{k} = kR \ , \ \bar{\omega} = \omega R/c \ .$$

Henceforth, nondimensional quantities are used, but for convenience the bars are omitted.

2.4 Nondimensional Form of Equations

The nondimensional forms of the elastic shell equations (2.1.1)
and (2.1.1) are

$$\beta \frac{\partial^2 \psi}{\partial x^2} + \frac{Q(1-\nu)}{2} (\frac{\partial w}{\partial x} - \psi) - \frac{\beta \gamma \alpha}{2\rho} \frac{\partial^2 \psi}{\partial t^2} = 0 \ , \qquad (2.4.1.)$$

$$w + \frac{Q(1-\nu)}{2} (\frac{\partial \psi}{\partial x} - \frac{\partial^2 w}{\partial x^2}) + \frac{\gamma \zeta}{2\rho} \frac{\partial^2 w}{\partial t^2} - \frac{p_w}{2} = 0 \ , \qquad (2.4.2)$$

and the nondimensional form of eq. (2.1.3) is

$$\beta \frac{\partial^4 w}{\partial x^4} + w = - \frac{\gamma \alpha}{2\rho} \frac{\partial^2 w}{\partial t^2} + \frac{\beta \gamma \alpha}{2\rho} \frac{\partial^4 w}{\partial x^2 \partial t^2} + \frac{p_w}{2} \ , \qquad (2.4.3)$$

where $\beta = \alpha^2/12$.

Nondimensional forms of the viscoelastic shell equations corresponding to eqs. (2.4.1) and (2.4.2) are not given since shear deformation is neglected in the analysis of the viscoelastic tube. The nondimensional form of the viscoelastic tube equation, with shear neglected,
is

$$2(\frac{\partial}{\partial t} + \frac{m}{\tau})(\beta \frac{\partial^4 w}{\partial t^2} + w) = (\frac{\partial}{\partial t} + \frac{1}{\tau})(p_w - \frac{\gamma \alpha}{\rho} \frac{\partial^2 w}{\partial t^2} + \frac{\gamma \alpha \beta}{\rho} \frac{\partial^4 w}{\partial x^2 \partial t^2}) \qquad (2.4.4)$$

for the incompressible standard material.

Nondimensional forms of the Euler equations are

$$\frac{\partial^2 u_x}{\partial t^2} = - \frac{\partial p}{\partial x} \ , \quad \frac{\partial^2 u_r}{\partial t^2} = - \frac{\partial p}{\partial r} \ , \qquad (2.4.5)$$

and the continuity equation is unchanged in form as in eq. (2.2.5).

2.5 Solution of Fluid Equations

A solution, in the form of a propagating wave travelling in the x direction, of

$$\frac{\partial^4 u_r}{\partial r^2 \partial t^2} - \frac{1}{r^2}\frac{\partial^4 u_r}{\partial t^2} + \frac{1}{r}\frac{\partial^3 u_r}{\partial r \partial t^2} + \frac{\partial^4 u_r}{\partial t^2 \partial x^2} = 0 , \qquad (2.5.1)$$

which is the nondimensional form of eq. (2.2.5), is sought. Consequently,

$$u_r(r,x,t) = \phi(r)\, e^{i(kx-\omega t)} ,$$

where k is the nondimensional wave number, ω is the nondimensional circular frequency, and i = $\sqrt{-1}$, is substituted in eq. (2.5.1) to obtain

$$\frac{d^2\phi}{dr^2} + \frac{1}{r}\frac{d\phi}{dr} - (\frac{1}{r^2} + k^2)\phi = 0 . \qquad (2.5.2)$$

Eq. (2.5.2) becomes

$$\frac{d^2\phi}{dz^2} + \frac{1}{z}\frac{d\phi}{dz} - (1 + \frac{1}{z^2})\phi = 0 , \qquad (2.5.3)$$

under the gauge transformation z = kr and the solution, which is regular at r = 0, is

$$\phi(r) = GI_1(kr) , \quad 0 \le r \le 1 ,$$

where I_1 is a modified Bessel function of the first kind and first order and G is at most a function of k. It follows that

$$u_r(r,x,t) = GI_1(kr) \, e^{i(kx-\omega t)} \, . \qquad (2.5.4)$$

More general solutions consist of terms of the form

$$u_r(r,x,t) = \int_{-\infty}^{\infty} G(k) \, I_1(kr) \, e^{i(kx-\omega t)} dk \, , \qquad (2.5.5)$$

which are obtained by superposition of elementary solutions of the type given by eq. (2.5.4).

Similarly it may be shown that general solutions of the nondimensional form

$$\frac{\partial^2 p}{\partial r^2} + \frac{1}{r} \frac{\partial p}{\partial r} + \frac{\partial^2 p}{\partial x^2} = 0 \, , \qquad (2.5.6)$$

of eq. (2.2.6) consist of terms of the form

$$p(r,x,t) = \int_{-\infty}^{\infty} H(k) \, I_0(ir) \, e^{i(kx-\omega t)} \, dk, \qquad (2.5.7)$$

where I_0 is a modified Bessel function of the first kind and order zero. The functions $G(k)$ and $H(k)$ in eqs. (2.5.5) and (2.5.7) are related through the second of eqs. (2.4.5) giving

$$\omega^2 G(k) = kH(k) \, . \qquad (2.5.8)$$

Substituting $r = 1$ in eqs. (2.5.5) and (2.5.7) gives

$$w = u_r(1,x,t) = \int_{-\infty}^{\infty} G(k) \; I_1(k) \; e^{i(kx-\omega t)} dk \; , \qquad (2.5.9)$$

$$p_i = p(1,x,t) = \int_{-\infty}^{\infty} H(k) \; I_0(k) \; e^{i(kx-\omega t)} dk \; . \qquad (2.5.10)$$

3. SOLUTION OF INITIAL VALUE PROBLEM

3.1 Dispersion relations for elastic tube

In order to obtain dispersion relations for the elastic tube when shear is taken into account the solution for ψ is taken to consist of terms of the form,

$$\psi = \int_{-\infty}^{\infty} F(k) \; I_1(k) \; e^{i(kx-\omega t)} \; dk \; . \qquad (3.1.1)$$

Substituting eqs. (3.1.1), (2.5.9) and (2.5.10) into eqs. (2.4.1) and (2.4.2), and employing the relation (2.5.8) gives

$$i \; \frac{Q(1-\nu)}{2} \; kG + [\omega^2 \xi - \frac{Q(1-\nu)}{2} \; \beta k^2] \; F = 0 \qquad (3.1.2)$$

$$[1 + \frac{Q(1-\nu)}{2} \; k^2 - \frac{\omega^2 \gamma \alpha}{2\rho} - \frac{\omega^2}{2k} \; I] \; G + i \; \frac{Q(1-\nu)}{2} \; kF = 0 \; , \qquad (3.1.3)$$

where

$$\xi = \frac{\gamma \alpha^3}{24\rho} \; , \; \beta = \frac{\alpha^2}{12} \; , \; I(k) = \frac{I_0(k)}{I_1(k)} \; .$$

In obtaining eqs. (3.1.2) and (3.1.3) it is assumed that the external perturbation pressure $p_e = 0$ for $t > 0$.

The dispersion relations are obtained by equating the coefficient determinant of F(k) and G(k) in eqs. (3.1.2) and (3.1.3) to zero and solving the resulting biquadratic equation in ω, which is

$$
\omega^4[\xi(\frac{k\gamma\alpha}{2\rho} + \frac{I}{2})] - \omega^2[\xi(k + \frac{Q(1-\nu)}{2} k^3)
$$
$$
+ (\frac{Q(1-\nu)}{2} + \beta k^2)(\frac{k\gamma\alpha}{2\rho} + \frac{I}{2})] + [\frac{Q(1-\nu)}{2} k
$$
$$
+ \beta k^3 + \frac{Q(1-\nu)}{2} \beta k^5] = 0 . \tag{3.1.4}
$$

The roots of eq. (3.1.4), which are real, are $\omega = \pm W_1(k)$, $\omega = \pm W_2(k)$, where $W_1 = (L-M)^{1/2}$, $W_2 = (L+M)^{1/2}$ and L and M are easily obtained from eq. (3.1.4). The contributions to the wave motion due to $W_1(k)$ and $W_2(k)$ are described as the first and second modes respectively. If the shear deformation is neglected, $\psi = \partial w/\partial x$, and there is only a single mode of propagation, the single shell equation in nondimensional form is given by eq. (2.4.3) and the corresponding dispersion relation is

$$
\omega = W(k) = \left[\frac{2k(1+ k^4)}{I + 2k(\frac{\gamma\alpha}{2\rho} + \xi k^2)} \right]^{1/2} . \tag{3.1.5}
$$

A realistic value of γ/ρ for an elastomer tube filled with a fluid whose density is close to that of water is $\gamma/\rho = 1.1$ and for biological applications a realistic value of α is $\alpha = 0.1$. The dispersion relations $W_1(k)$ and $W_2(k)$ are shown graphically in Fig. 3.1.1, for $\alpha/\rho = 1.1$, $\nu = 0.5$, $\alpha = 0.05$ and $\alpha = 0.1$, along with the dispersion relation (3.1.5). It is evident that the second mode has a cut-off frequency below which a wave does not propagate.

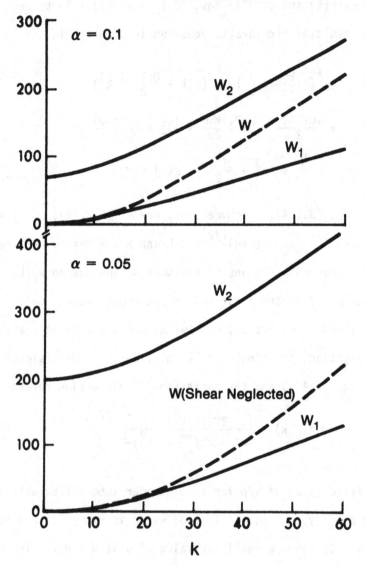

Fig. 3.1.1 Dispersion relations

The phase speeds for the first and second modes are given by

$$c_1 = W_1(k)/k \text{ and } c_2 = W_2(k)/k \text{ ,}$$

respectively, and the group speeds by

$$C_1 = \frac{dW_1(k)}{dk} \quad \text{and} \quad C_2 = \frac{dW_2(k)}{dk} \, .$$

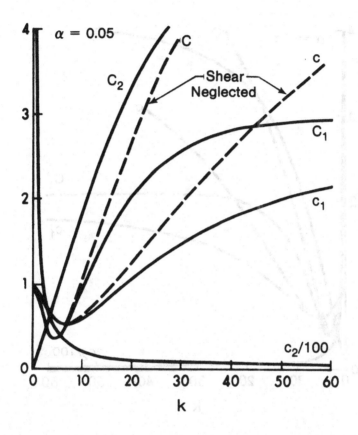

Fig. 3.1.2   Phase and group speeds for α = 0.0.5

These are shown in Figs. 3.1.2 and 3.1.3 along with phase and group speeds obtained from eq. (3.1.5). The results shown in Figs. (3.1.1), (3.1.2) and (3.1.3) indicate that the dispersion relations and group and phase speeds obtained by neglecting shear are good approximations to those of the first mode for low wave numbers and, as expected, that the lower the value of α = h/R, the greater the justification for neglecting shear deformation of the tube wall. When an initial value problem is considered the quality of the approximation obtained by neglecting shear depends on the initial conditions as well as on α.

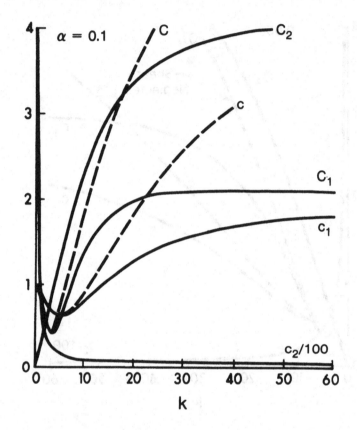

Fig. 3.1.3 Phase and group speeds for $\alpha = 0.1$

3.2 Effect of Shear Deformation on Solution

The ratios $|F_1/G_1|$ and $|F_2/G_2|$, where the subscripts refer to the first and second modes, are known as the mode shapes and are obtained by substituting $\{W_1(k)\}^2$ and $\{W_2(k)\}^2$, respectively for ω^2 in eqs. (3.1.2) or (3.1.3), to give

$$\frac{F_n}{G_n} = iS_n \ , \ n = 1,2 \tag{3.2.1}$$

where
$$S_n = k \left[\frac{2(\beta k^2 - W_n^2 \xi)}{Q(1-\nu)} + 1 \right]^{-1} \tag{3.2.2}$$

In general a wave motion is composed of a superposition of the two modes and this can be illustrated by considering an initial value problem with the system quiescent at t = 0, so that the initial conditions are of the form

$$w(x,0) = g(x) \ , \ \psi(x,0) = f(x) \ , \tag{3.2.3}$$

$$\dot{w}(x,0) = 0 \ , \ \dot{\psi}(x,0) = 0 \ . \tag{3.2.4}$$

Because of initial conditions (3.2.4) the solution can be expressed in the form

$$w = \sum_{n=1}^{2} \int_{-\infty}^{\infty} G_n(k) [e^{i(kx - W_n(k)t)} + e^{i(kx + W_n(k)t}] \ dk \ , \tag{3.2.5}$$

$$\psi = \sum_{n=1}^{2} \int_{-\infty}^{\infty} F_n(k) [e^{i(kx - W_n(k)t)} + e^{i(kx + W_n(k)t)}] dk \tag{3.2.6}$$

and it then follows from initial conditions (3.2.3) that

$$g(x) = 2 \sum_{n=1}^{2} \int_{-\infty}^{\infty} G_n(k) \ e^{ikx} dk \ , \ f(x) = 2 \sum_{n=1}^{2} \int_{\infty}^{\infty} F_n(k) e^{ikx} dk. \tag{3.2.7}$$

The functions f(x) and g(x) can also be expressed in the form

$$g(x) = \int_{-\infty}^{\infty} G(k) e^{ikx} dk \ , \ f(x) = \int_{-\infty}^{\infty} F(k) \ e^{ikx} dk \ , \tag{3.2.8}$$

where $G(k) = \frac{1}{2\pi} \int_{-\infty}^{\infty} g(x) \ e^{-kx} \ dx \ , \ F(k) = \frac{1}{2\pi} \int_{-\infty}^{\infty} f(x) \ e^{-kx} \ dx \ . \tag{3.2.9}$

Comparing eqs. (3.27) and (3.28) we obtain

$$2[G_1(k) + G_2(k)] = G(k) \tag{3.2.10}$$

$$2[F_1(k) + F_2(k)] = F(k) . \tag{3.2.11}$$

Eqs. (3.2.9) are used to obtain $G(k)$ and $F(k)$ and eqs. (3.2.1), (3.2.10) and (3.2.11) provide four equations which can readily be solved to obtain the four unknowns $G_n(k)$ and $F_n(k)$, $n = 1,2$, which are in turn substituted in eqs. (3.2.5) and (3.2.6) to obtain a formal solution to the initial value problem. The integrals in eqs. (3.2.5) and (3.2.6) can be evaluated numerically or otherwise.

In order to obtain the functions $w(x,0)$ and $\psi(x,0)$ given by initial conditions (3.2.3) the static form of eqs. (2.4.1) and (2.4.2) with $p_w = - p_e$,

$$\beta \frac{d^2\psi}{dx^2} + \frac{Q(1-\nu)}{2} (\frac{dw}{dx} - \psi) = 0 , \tag{3.2.12}$$

$$w + \frac{Q(1-\nu)}{2} (\frac{d\psi}{dx} - \frac{d^2w}{dx^2}) + \frac{q\delta}{2} (x) = 0 , \tag{3.2.13}$$

must be solved. In Eqs. (3.2.12) and (3.2.13), p_e is given by eq. (1.2.1) with $t < 0$.

Applying standard Fourier transform techniques to equations (3.2.12) and (3.2.13) gives

$$G(k) = - \frac{q}{4\pi} \left[\frac{\beta k^2 + \kappa}{\kappa\beta k^4 + k^2\beta + \kappa} \right] , \tag{3.2.14}$$

$$F(k) = -\frac{iq\kappa}{4\pi}\left[\frac{k}{\beta k^4 + k^2\beta + \kappa}\right] \tag{3.2.15}$$

where $G(k)$ and $F(k)$ are the functions defined by eq. (3.2.4) and $\kappa = Q(1-\nu)/2$. The Fourier transforms (3.2.14) and (3.2.15) can be inverted according to eqs. (3.2.7) to give

$$g(x) = -\frac{q}{4}\beta^{-1/4}\cos ec2\theta \exp(-\beta^{1/4}|x|\cos\theta)[2\cos 2\theta \sin(\theta-\beta^{-1/4}|x|\sin\theta)$$

$$+ \sin(\theta+\beta^{-1/4}|x|\sin\theta)] \tag{3.2.16}$$

$$f(x) = \frac{q}{4}\beta^{1/2}\cos ec2\theta \exp(-\beta^{-1/4}|x|\cos\theta)\sin(\beta^{-1/4}|x|\sin\theta) \tag{3.2.17}$$

where $\cos 2\theta = \beta^{1/2}/(2\kappa)$. If shear deformation of the tube wall is neglected the static displacement is obtained from the static equivalent of eq. (2.4.3) again with $p_w = -q\delta(x)$, and is

$$w(x,0) = g(x) = \frac{-q}{2^{3/2}\beta^{1/4}}\exp(-2^{1/2}\beta^{-1/4}|x|)\{\cos(2^{-1/2}\beta^{-1/4}|x|)$$

$$+ \sin(2^{-1/2}\beta^{-1/4}|x|)\} \tag{3.2.18}$$

and the Fourier transform defined by the first of equations (3.2.9), is denoted by $\hat{G}(k)$ and is given by

$$\hat{G}(k) = -\frac{q}{4\pi}\left(\frac{1}{\beta k^4 + 1}\right). \tag{3.2.19}$$

In Fig. 3.2.1 we show the initial displacements $w(x,0) = g(x)$, for $\alpha = 0.05$ and $\alpha = 0.1$, given by eq. (3.2.16) with $q = 1$. As before, numerical results are presented for the parameters $\gamma/\rho = 1.1$, $\nu = 0.5$,

Q = 0.91 and for q = 1.

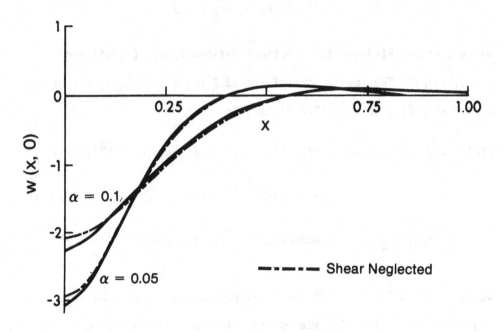

Fig. 3.2.1 Initial static displacement due
to external pressure p_e = $q\delta(x)$ with q = 1

Eqs. (3.2.1), (3.2.2), (3.2.10) and (3.2.11) are used along with
(3.2.14) and (3.2.15) to obtain $G_1(k)$ and $G_2(k)$. The contributions
$G_1(k)$ and $G_2(k)$, of the first and second modes, respectively, to the
tube wall displacement w, for a given wave number k, are shown graphi-.
cally in Figs. 3.2.2 and 3.2.3 along with $\hat{G}(k)/2$ given by equation
(3.2.19). If the effect of shear on the displacement w of the tube
wall is negligible, then $W_1(k) \simeq W(k)$, $G_1(k) >> G_2(k)$, so that $G_1(k) \simeq$
$G(k)/2 \simeq \hat{G}(k)/2$ for the range of k outside of which $G_1(k)$ and $G_2(k)$
are negligible. The results shown in the figures indicate that for
the initial value problem considered the effect of shear deformation
is negligible for α = 0.05. For α = 0.1 the effect of shear deforma-

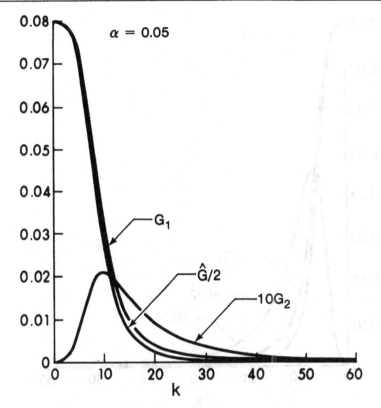

Fig. 3.2.2 Contributions $G_1(k)$ and $G_2(k)$ to the
radial displacement w for $\alpha = 0.05$ and $q = 1$

tion is more significant but still small enough that its neglect may

be justified. If $\alpha > 0.1$ then shear deformation of the tube wall

should be included in the analysis. In the following sections it is

assumed that neglect of shear deformation is justified in the analysis.

3.3 Solution for Viscoelastic Tube

In this section the solution, for a viscoelastic tube composed of

the incompressible standard material, is obtained. The constitutive

equation is given by eq. (1.3.3) and the nondimensional governing

equation is eq. (2.4.4). Shear deformation is neglected and the elastic

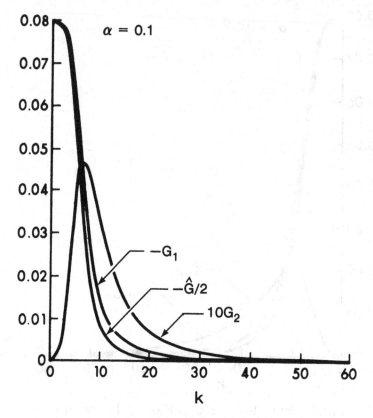

Fig. 3.2.3 Contributions $G_1(k)$ and $G_2(k)$ to the
radial displacement w for $\alpha = 0.1$ and $q = 1$

solution is obtained as a limiting case as $m \to 1$.

It is assumed that the circumferential line loading

$$p_e = q\delta(x) [1-H(t)] , \qquad (3.3.1)$$

is applied at time $t = -\infty$ so that at time $t = 0$, w and all the time derivatives of w are zero. It follows then that the initial deflected form is given by the solution of

$$\frac{d^4 w}{dx^4} (x,0) + w(x,0) = - \frac{q}{2m} \delta(x) . \qquad (3.3.2)$$

Eq. (3.3.2) is the same as that for an elastic tube with shear modulus equal to the relaxed modulus of the viscoelastic tube. The deflected form $w(x,0)$ at $t = 0$, obtained from eq. (3.3.2) is given by

$$w(x,0)/w_0 = \exp(-2^{-1/2}\beta^{-1/4}|x|)\{\cos(2^{-1/2}\beta^{-1/4}|x|)+ \sin(2^{-1/2}\beta^{-1/4}|x|)\}$$

$$(3.3.3)$$

where $w_0 = w(0,0)$.

The nondimensional intensity of line loading is given by

$$q = - 2^{3/2}\beta^{1/4} m w_0 .$$

Since the system is initially at rest eq. (3.3.2) and

$$\dot{w}(x,0^+) = 0$$

provide two initial conditions. A third initial condition is required, but this is not obtained explicitly, but is obtained in the Fourier transform domain.

We adopt the following form for the Fourier transform $\tilde{f}(k)$ of a function $f(x)$,

$$\tilde{f}(k) = \frac{1}{2\pi} \int_{-\infty}^{\infty} f(x)\, e^{-ikx}\, dx ,\qquad (3.3.4)$$

so that

$$f(x) = \int_{-\infty}^{\infty} \tilde{f}(k)\, e^{ikx}\, dk .\qquad (3.3.5)$$

Eqs. (3.3.4) and (3.3.5) are consistent with eqs. (3.2.8) and (3.2.9). Henceforth a superposed tilde denotes the Fourier transform as defined by eq. (3.3.4).

The initial value problem considered involves temporal attenuation rather than spatial attentuation of the perturbation. Consequently we express the radial deflection $w(x,t)$ and pressure $p(r,x,t)$ as Fourier integrals

$$w(x,t) = \int_{-\infty}^{\infty} \tilde{w}(k,t) \, e^{ikx} \, dk \, , \tag{3.3.6}$$

$$p(r,x,t) = \int_{-\infty}^{\infty} \tilde{p}(r,k,t) \, e^{ikx} \, dk \, . \tag{3.3.7}$$

The Fourier transform of eq. (2.5.6) is

$$\frac{d^2\tilde{p}}{dr^2} + \frac{1}{r} \frac{d\tilde{p}}{dr} - k^2\tilde{p} = 0 \, , \tag{3.3.8}$$

so that $$\tilde{p}(r,k,t) = \tilde{p}_0(k,t) I_0(kr) \, . \tag{3.3.9}$$

It follows from the Fourier transform of the second of eqs. (2.4.5) and eq. (3.3.9) that

$$\frac{\partial^2 \tilde{u}_r}{\partial t^2} = - \tilde{p}_0(k,t) \, kI_1(kr) \, , \tag{3.3.10}$$

and for $r = 1$ this gives

$$\frac{\partial^2 \tilde{w}}{\partial t^2} = - \tilde{p}_0(k,t) \, kI_1(k) \, . \tag{3.3.11}$$

It follows from eq. (3.3.9) that

$$\tilde{p}_i = \tilde{p}_0(k,t) \, I_0(k) \, , \tag{3.3.12}$$

and eliminating $p_0(k,t)$ from eqs. (3.3.11) and (3.3.12) gives

$$\frac{\partial^2 w}{\partial t^2} = - \frac{kI_1(k)}{I_0(k)} \tilde{p}_i(k,t) . \tag{3.3.13}$$

We multiply the Fourier transform of eq. (2.4.4) by $kI_1(k)$ and use eq. (3.3.13) to obtain

$$[I_0(k) + \frac{\gamma\alpha}{\rho} (\beta k^2+1) kI_1(k)] \frac{\partial^3 \tilde{w}}{\partial t^3}$$

$$+ \frac{1}{\tau} [I_0(k) + \frac{\gamma\alpha}{\rho} (\beta k^2+1) kI_1(k)] \frac{\partial^2 \tilde{w}}{\partial t^2}$$

$$+ 2kI_1(k)(\beta k^4+1) \frac{\partial \tilde{w}}{\partial t}$$

$$+ \frac{2mk}{\tau} I_1(k)(\beta k^4+1) \tilde{w} = - kI_1(k)[\frac{p_e}{\tau} + \frac{\partial p_e}{\partial t}] , \tag{3.3.14}$$

where $\qquad\qquad \tilde{p}_e = \frac{q}{2\pi} [1 - H(t)] .$

Then, by integrating this equation over $-\varepsilon \leq t \leq \varepsilon$ and, letting $\varepsilon \to 0$, we obtain $w(k,0^+) = w(k,0^-)$, $\dot{\tilde{w}}(k,0^+) = \dot{\tilde{w}}(k,0^-) = 0$ and

$$\tilde{w}(k,0^+)/w_0 = \frac{-2^{3/2} m\beta^{1/4} kI_1(k)}{\pi[I_0(k) + \frac{\gamma\alpha k}{\rho} (1+\beta k^2) I_1(k)]} \tag{3.3.15}$$

For $t > 0$ we can write eq. (3.3.14) as

$$\frac{\partial^3 \tilde{w}}{\partial t^3} + \frac{1}{\tau} \frac{\partial^2 \tilde{w}}{\partial t^2} + \Lambda \frac{\partial \tilde{w}}{\partial t} + \frac{m}{\tau} \Lambda \tilde{w} = 0 \tag{3.3.16}$$

and the initial conditions are

$$\tilde{w}(k,0^+)/w_0 = \frac{2^{1/2}\beta^{1/4}}{\pi(\beta k^4+1)} \quad , \tag{3.3.17}$$

$$\dot{\tilde{w}}(k,0^+) = 0 \quad , \tag{3.3.18}$$

and eq. (3.3.15). In eq. (3.3.16),

$$\Lambda = \frac{2kI_1(k)(1+\beta k^4)}{I_0(k) + \frac{\gamma\alpha}{\rho}(1+\beta k^2) \, kI_1(k)} \quad .$$

Substituting $\tilde{w} = Be^{\phi t}$, where B is a constant, in eq. (3.3.16) gives the cubic equation

$$\phi^3 + \frac{1}{\tau}\phi^2 + \Lambda\phi + \frac{m}{\tau}\Lambda = 0 \quad . \tag{3.3.19}$$

When $0 < m < 1$ it is easily shown that the real parts of the roots will be negative for all real k as they must be from purely physical considerations. The condition for one real root and two complex conjugate roots is

$$\frac{b^2}{4} + \frac{a^3}{27} > 0$$

where
$$a = \frac{1}{3}(3\Lambda - \frac{1}{\tau^2}),$$

and
$$b = \frac{1}{27\tau}(27m\Lambda - 9\Lambda + \frac{2}{\tau^2}) \quad .$$

It may then be shown that if $\frac{1}{9} \le m < 1$, eq. (3.3.19) has precisely one real negative root and two complex conjugate roots for all real k

and $\tau > 0$ and we consider this case. The real root of (3.3.19) is then

$$\phi_1 = [\{-\frac{b}{2} + (\frac{b^2}{4} + \frac{a^3}{27})^{1/2}\}^{1/3} + \{-\frac{b}{2} - (\frac{b^2}{4} + \frac{a^3}{27})^{1/2}\}^{1/3}] - \frac{1}{3\tau} \quad (3.3.20)$$

and the complex conjugate roots are

$$\phi_2 = -\frac{1}{2}(\frac{1}{\tau} + \phi_1) + \frac{i\gamma}{2}, \qquad (3.3.21)$$

$$\phi_3 = -\frac{1}{2}(\frac{1}{\tau} + \phi_1) - \frac{i\gamma}{2}, \qquad (3.3.22)$$

where γ is the positive square root of

$$\gamma^2 = 4\Lambda - (\frac{1}{\tau} + \phi_1)(\frac{1}{\tau} - 3\phi_1) .$$

Putting the roots in the form

$$\phi_1 = -X_1(k) , \quad \phi_{2,3} = -X_2(k) \pm i\omega(k) ,$$

X_1, X_2, and ω are positive and functions of k which can be deduced

from eqs. (3.3.20), (3.3.21) and (3.3.22). The solution of eq.

(3.3.15) is then given by

$$\tilde{w}(k,t)/w_0 = A_1 e^{-X_1 t} + e^{-X_2 t}(A_2\cos\omega t + A_3 \sin\omega t) . \quad (3.3.23)$$

Using the initial conditions (3.3.15), (3.3.17) and (3.3.18) it is a

routine calculation to obtain A_1, A_2 and A_3 for a given value of k.

The solution of the governing equation (2.4.4), with p_e given by

eq. (3.3.1), is then given formally by

$$w(x,t)/w_0 = \int_{-\infty}^{\infty} [A_1 e^{-X_1 t} + e^{-X_2 t}(A_2\cos\omega t + A_3\sin\omega t)] \, e^{ikx} \, dk \quad .(3.3.24)$$

It may be shown that A_1, A_2, X_1, X_2, and ω are even functions of k while A_3 is an odd function.

Evaluation of the Fourier integral (3.3.24), that is inversion of the Fourier transform (3.3.23) appears to be intractable analytically, consequently results have been obtained numerically applying the Fast Fourier Transform algorithm, as described by Brigham (1974).

Since the Fourier transform given in eq. (3.3.23) is an even function of k the Fourier integral (3.3.24) can be rewritten as

$$w(x,t)/w_0 = 2 \int_{0}^{\infty} [A_1 e^{-X_1 t} + e^{-X_2 t}(A_2\cos\omega t + A_3\sin\omega t)] \, \cos kx \, dk.$$

This form is more convenient for numerical evaluation.

Results are shown graphically in terms of nondimensional quantities in Figs. 3.3.2 to 3.3.5, for $\gamma/\rho = 1.1$, which is a realistic value for an elastomer tube filled with biological fluid, and $\alpha = 0.05$. The initial displacement $w(x,0)/w_0$ is shown in Fig. 3.3.1. In Figs. 3.3.2 and 3.3.3 $w(x,t_1)/w_0$ is shown for $t_1 = 0.5$, 1.0, 2.0 and $m = 0.5$. Two different relaxations times, $\tau = 1.0$ and $\tau = 10$ are considered. In Fig. 3.3.4 the results for an elastic tube are shown and they can be compared with those for $m = 0.9$ and $\tau = 1$ shown in Fig. 3.3.5.

As $m \rightarrow 1$ the standard viscoelastic model approached that for an elastic material and this is indicated by the results shown in Figs. 3.3.4 and 3.3.5. The results for the elastic tube we obtained by

putting m = 1 in the program for the viscoelastic tube. It is clear
that the influence of m is greater than of the relaxation time τ.

For biological applications m = 0.5 is a realistic value and the
response of the tube wall is a rapidly damped dispersed wave form,

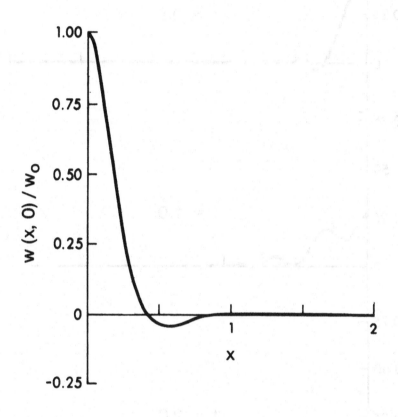

Fig. 3.3.1   Initial displacement

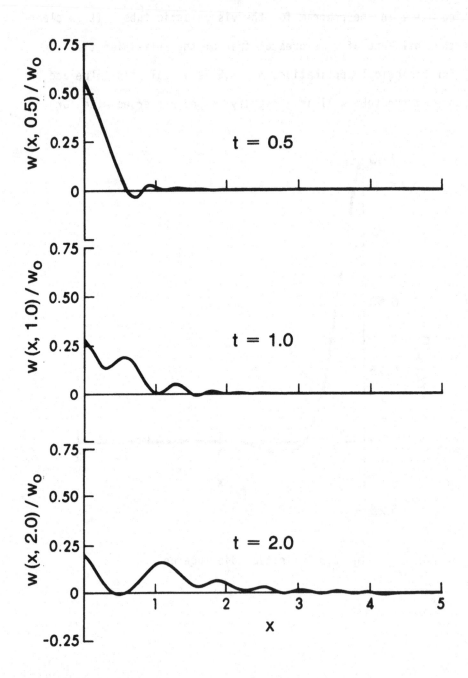

Fig. 3.3.2 $w(x,t)/w_0$ for m = 0.5, τ = 1, α = 0.05

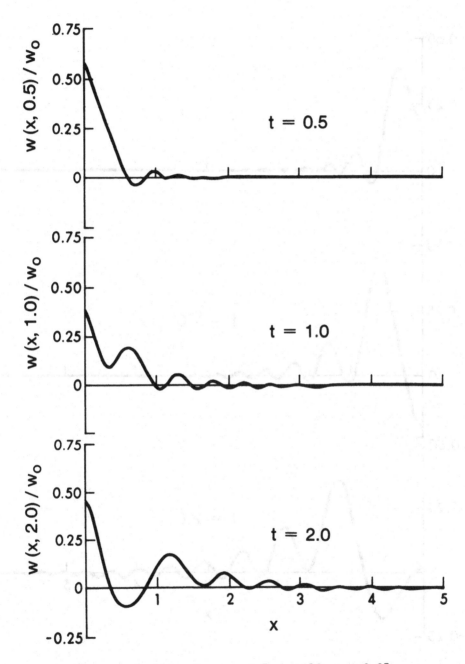

Fig. 3.3.3 $w(x,t)/w_0$ for $m = 0.5$, $\tau = 10$, $\alpha = 0.05$

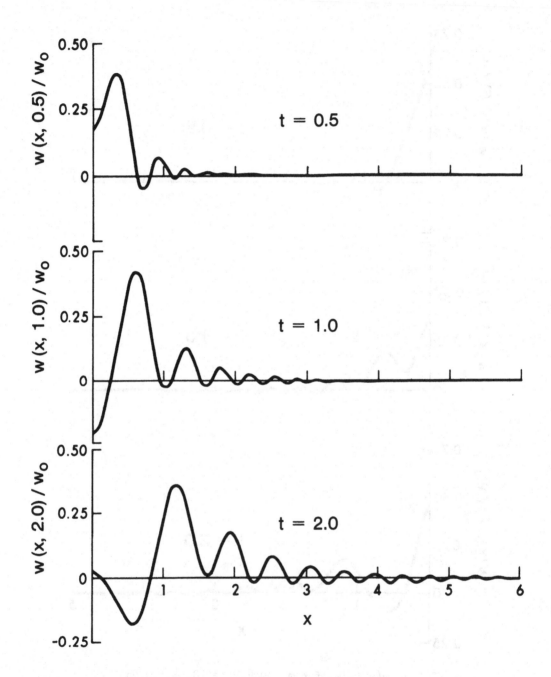

Fig. 3.3.4 $w(x,t)/w_0$ for elastic tube, $\alpha = 0.05$

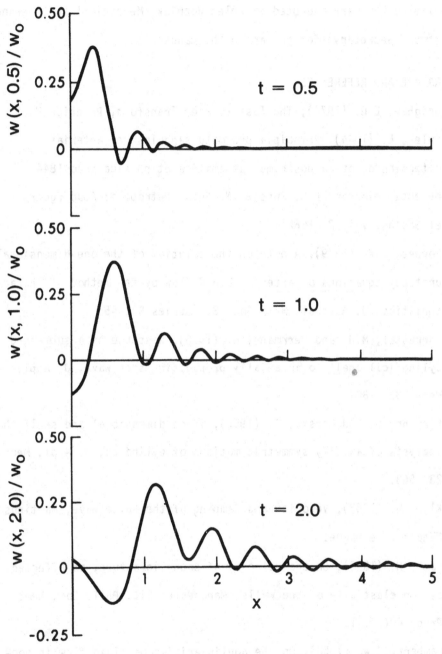

Fig. 3.3.5 $w(x,t)/w_0$ for m = 0.9, τ = 1, α = 0.05

ACKNOWLEDGEMENT

The authors are indebted to Helen Wozniuk (Mechanical Engineering Department Secretary) for preparing the manuscript.

LITERATURE AND REFERENCES

1. Brigham, E.O. (1974), The Fast Fourier Transform, Prentice Hall.

2. Euler, L. (1844), Principia pro motu sanguins per arterias determinado, opera posthuma mathematica et physica anno 1844 detecta, eiderant P.H. Fuss et N. Fuss. Petropoli: Apud Eggers et Socios, Vol. 2, 1862.

3. Forbes, L.K. (1979), A note on the solution of the one-dimensional unsteady equations of arterial blood flow by the method of charac- teristics, J. Austral. Math Soc. 21 (Series B), 45.

4. Forrestal, M.J. and Herrmann, G. (1965), Response of a submerged cylindrical shell to an axially propagating step wave, J. Appl. Mech. 32, 788.

5. Herrmann, G. and Mirsky, I. (1956), Three dimensional and shell theory analysis of axially symmetric motions of cylinders, J. Appl. Mech. 23, 563.

6. Klip, W. (1962), Velocity and Damping of the Pulse Wave, Martinus Nÿhoff, The Hague.

7. Lamb, H. (1898), On the velocity of sound in a tube, as affected by the elasticity of the walls, Manchester Lit. Phil. Soc. Mem. Proc. 42(9), 1.

8. Lambert, J.W. (1958), On the nonlinearities of fluid flow in non- rigid tubes, J. Franklin Inst., 266, 83.

9. Rubinow, S.I. and Keller, J.B. (1968), Hydrodynamic aspects of the
    circulatory system, In Hemorheology, Proc. 1st. Int. Conf. Univ.
    of Iceland, A.L. Copley, Ed., Pergamon.

10. Rubinow, S.I. and Keller, J.B. (1971), Wave propagation in a fluid
    filled tube, J. Acoust. Soc. Am. 50, 198.

11. Rubinow, S.I. and Keller, J.B. (1978), Wave propagation in a visco-
    elastic tube containing a viscous fluid, J. Fluid Mech. 88, 181.

12. Rudinger, G. (1966), Review of current mathematical methods for
    the analysis of blood flow, Biomedical Fluids Symposium, ASME,
    New York.

13. Rudinger, G. (1970), Shock waves in mathematicsl models of the
    aorta, J. Appl. Mech. 37, 34.

14. Skalak, R. (1966), Wave propagation in blood flow in biomechanics,
    Proc. Symp. Appl. Mech. Div. ASME, Y.C. Fung, Ed.

15. Tait, R.J., Moodie, T.B. and Haddow, J.B. (1981), Wave propagation
    in a fluid-filled elastic tube, Acta Mechanica, 38, 71.

16. Teipel, I. (1973), Michtlineare wellenausbreitungsvorgänge in
    elastischen leitungen, Acta Mechanica, 16, 93.

17. Witzig, K. (1914), Uber erzwungene Wellenbewegungen zäher,
    inkompressibler Flussigkeiten in elastischen Röhren, Inaugural
    Dissertation, Universitat Bern, K.J. Wyss, Bern.

CHAPTER IV
SMALL ARTERIES AND THE INTERACTION WITH
THE CARDIOVASCULAR SYSTEM

THOMAS   KENNER
Physiologisches Institut
Universität Graz
Graz, Austria

1.   INTRODUCTION

The purpose of this paper is the presentation and
discussion of problems which are related to the reactions
of small resistance vessels and their upstream and downstream
effects. It is intended to give a short and broad overlook
over this field. The function of small arteries cannot be
understood without considering the interaction with other
parts of the cardiovascular system. These small muscular
vessels are executing control simultaneously on local blood
flow and on arterial blood pressure. This dual task sometimes leads to contradicting trends; e.g. during physical
exercise we find vasodilatation in order to increase local
blood flow, and at the same time vasoconstriction in order

to prevent breakdown of the arterial blood pressure. Hydrodynamic phenomena are closely intertwined with metabolically
or neurally mediated smooth muscle reactions.

In this article many references are made to the work of
E.Wetterer and his pupils, and to the book by Wetterer and
Kenner (1968). More recent work from our group has been re-
viewed during a Satellite Symposium to the World Congress of
Physiology in Budapest (Kenner et al. 1982).

The work described in this article has been supported
by the Austrian Science Research Fund.

1.1 What is a small artery ?

Before I was invited to discuss blood flow in small
arteries I certainly did not realize the difficulty to define the term "small artery". An agreement on the gross
anatomical nomenclature is found for all vessels except just
the small arteries which, in some sources (e.g. Noordergraaf
1978) may be listed under the name "terminal arteries". In
other sources terminal arteries and small arteries are
assumed as different entities. In general, the successive
anatomical orders of arteries are called: aorta, large
arteries, main arterial branches, terminal branches, small
arteries and arterioles. From the arterioles rise the capillaries and thence the venoles and the venous part of the
system.

Schmid-Schönbein (1976) cites a paper by Mall who in 1888 examined the structure of the canine mesenteric vasculature and points to the quite important fact that "Malls data are quoted in practically all textbooks of physiology throughout the world (often as treated by Green (1944)). Important physiological concepts.....are based on these archaic findings."

Noordergraaf (1978) contrasts in a table Greens compiled canine data with those by Wiedeman (1962,1963) who measured and counted microvessels in the bat wing. Noordergraafs table lists terminal arteries as having a diameter of 600 μm according to Green and 19 μm according to Wiedeman. The corresponding diameters of arterioles are 20 μm (Green), 7 μm (Wiedeman), and those of capillaries 8 μm (Green) and 3.7 μm (Wiedeman).

The whole problem seems to be complicated because of large differences in different vascular beds. In any case it seems to me necessary to point out that very often useful and good data are missing, particularly for the most basic and obvious phenomena, because nobody realizes that such data have not been carefully measured with appropriate methods.

Fortunately in most vascular areas reasonably similar vascular pattern can be found. One example of the terminal bed of the bat wing is shown in fig.1 (Mayrovitz et al.1976). This figure demonstrates how small arteries give rise to arterioles and how several capillaries branch off from one

arteriole. In many publications on arterial flow and especially in textbooks a dichotomic branching scheme is ex-
plicitely or implicitely assumed. Actually, dichotomic branching plays an important role in single branches of larger
or medium size arteries.

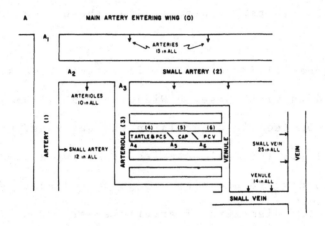

Fig.1 Topological model of arterial branching
(Mayrovitz et al.1976)

A more general scheme of a series - parallel network model
is shown in fig.2

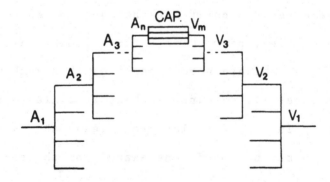

Fig.2 Series - parallel network (Popel 1980)
A arterioles,CAP capillaries,V venoles

From a functional viewpoint Lee and Nellis (1974) could show
that the microcirculatory transit times can be modeled best
by an array of vessels as shown in fig.3

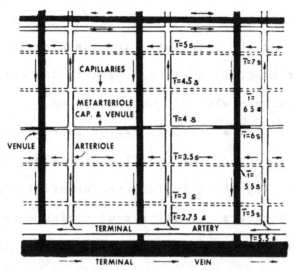

Fig. 3 Distribution of mean transit times in
a vascular network (Lee and Nellis 1974)

The schematic outlines of microvascular beds shown in
the preceding figures represent the three possible viewpoints
of presentation: fig.1 shows the anatomical viewpoint which
is valid for one particular example. Fig.2 is a more general
scheme, compromising the exact and the restricted view with
an attempt of generalization. Fig. 3 represents a theoretical
simplification of an actually rather more complex three dimensional network (Metzger 1973, Popel 1980).

It seems that we can describe the microcirculation in
most general terms as consisting of structural "modules" of
vessels the pattern of which is repeated in a certain tissue.
Fig. 3 represents such a repeating pattern. Another morpho-

logic pattern of such microvascular moduli was described by
Lipowsky and Zweifach (1974).

In conclusion we can define the region of "small arteries" as extending between larger and medium size arteries
on one hand and the arterioles and the microvascular moduli
on the other hand. We therefore have to include vessels with
a diameter between about 20 μm and 100 μm.

In a functional sense this region is essential for the
control and biological variability of the peripheral resistance, especially if we include the arterioles and the precapillary sphincters - small muscular ring structures around
the entrance of true capillaries. Capillaries themselves do
not contain contractile elements. In other words, if we discuss the function of the small arteries we must not exclude
the downstream part of the microcirculation.

1.2 Shear rate and shear stress and resistance to flow

Assuming laminar Poiseuille flow the pressure gradient
along a small artery is related to the flow q and the average
velocity $\overline{v}$ by

$$q = \overline{v} \, r^2 \pi \tag{1}$$

and

$$\Delta p / \Delta x = 8 \eta \, \overline{v} / r^2 \tag{2}$$

Since the shear stress τ_s at the vessel wall is proportional
to the shear rate dv/dr by

$$\tau_s = \eta \; dv/dr \tag{3}$$

and since in laminar flow

$$v = (r_i^2 - r^2) \; \Delta p / (4 \eta \Delta x) \tag{4}$$

(r_i is the radius at the vessel wall), we find the following
relations for the shear stress at the radius r:

$$\tau_s = - (\Delta p / \Delta x)(r/2) \tag{5}$$

or

$$\tau_s = - 4 \eta \bar{v}/r \tag{6}$$

or

$$\tau_s = - 4 \eta \; q/(r^3 \pi) \tag{7}$$

The values of the shear rates at the vessel wall r_i are in
the order of 100 to 200 s^{-1} in the aorta and are found to
rise slightly towards smaller vessels. The largest shear
rates are reported for capillaries. Schmid-Schönbein (1976)
cites values of about 300 s^{-1}. The corresponding shear
stresses are quite small and range between less than 1 to 10
Pa. (1 Pa corresonds to 10 μbar = 10 dyn/cm^2). Data about
values of shear rates and shear stresses in vivo differ in
different sources (e.g Whitmore 1968 and Caro et al. 1978
report shear rates in capillaries up to 1000 s^{-1}). Shear
stresses in the normal circulation, in any case, are 3 to 4
magnitudes smaller than the values of the arterial pressure
(mean value 13.3 kPa = 133 mbar) and - as will be discussed
in section 2.2 - 4 to 5 magnitudes smaller than the circumferential stress in the arterial wall. As Patel et al. find,
shear stress values above 40 Pa = 400 ubar may damage the

vascular endothelium.

The resistance to flow of one vessel segment of length Δx equals

$$R = \Delta p / q \tag{8}$$

and, therefore

$$R = 8 \eta \Delta x / (r^4 \pi) \tag{9}$$

The total resistance of n equal parallel vessels of resistance R_i is

$$R_{total} = R_i / n \tag{10}$$

The "hydraulic hindrance" $= \Delta x / ((n.2r)^4) \tag{11}$

applied by Schmid-Schönbein (1976) is a magnitude proportional to the resistance R_{total} of n parallel vessels:

"hydraulic hindrance" $= R_{total} \pi / (128 \eta) \tag{12}$

In this short summary of simplified equations it should also be mentioned shortly that blood shows non-Newtonian behaviour: the relation between shear rate and shear stress cannot be described by the linear relation eq.3. Several attempts to describe this relation have been made. Two equations cited from Noordergraaf (1978):

$$\tau_s = b(dv/dr)^s + C \tag{13}$$

where C is the yield stress necessary to initiate flow. b and s are constants. The following Casson equation describes the behaviour of blood in a wide range of shear rates:

$$\tau_s^{1/2} = k(dv/dr)^{1/2} + c^{1/2} \tag{14}$$

1.3 The arterial tree

According to Vadot (1967) 9 orders (0 to 8) of vessel sizes can be distinguished from the aorta to the capillaries. As mentioned above, in many single arterial branches, especially in the extremities, in the mesentery and in the lung, the main type of branching appears to be dichotomic. However, the overall count of vessels of each order shows that from each parent level about 15 daughter vessels are branching off.

If z is the order number of each level, n_z , D_z and l_z are number, diameter and length, respectively, of each level, then the following model relations can be assumed:

$$n_z = n_o k_n{}^z \qquad (1)$$

$$D_z = D_o k_D{}^z \qquad (2)$$

$$l_z = l_o k_l{}^z \qquad (3)$$

k are dimensionless constants describing the behaviour of each variable from level to level.

Fig.4 after Vadot (1967) shows the distribution of vessels as calculated for a 60 kg adult with an assumed resting cardiac output of 4.5 l/min. The corresponding equations of this concrete example are as follows:

$$n_z = 1 \ (15.4)^z \qquad (4)$$

$$D_z = 2 \ (0.376)^z \qquad (5)$$

$$l_z = 59 \ (0.475)^z \qquad (6)$$

with z = 0,1,2,.....8.

The "hydraulic hindrance" of this system is

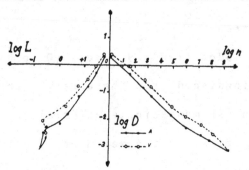

Fig. 4 Relation between log n (left, n: number of vessels), log L (right, L length of vessel segment) and log D (ordinate, D: diameter of the vessel) in the human circulation. The plot can be modeled by eq. 4 to eq.6. Closed circles: arteries. Open circles : veins. After Vadot (1967)

$$\text{"hydraulic hindrance"} = l_z / (n_z D_z^4) = 3.69 \, (1.56)^z \quad (7)$$

The highest value of the resistance to flow, therefore, is found in the capillary level (level 8 according to the model). The resistance of the small arteries and arterioles (levels 5 to 7) corresponds to 48% of the total peripheral resistance of the model whereas the estimated resistance of the capillary bed alone makes 37% of the total. The small rest of 15% is due to the resistance of medium and larger arteries (levels 0 to 4). - It is important to note that these values may be approximately valid for resting state. We can expect marked biological variability and large changes during exercise (see section 4.5).

The shear rate as well as the shear stress are proportional to

$$\tau_s \backsim D_z^{-3} n_z^{-1} \quad (8)$$

Insertion of the corresponding numerical value of the model

assumption leads to

$$\tau_s \hookrightarrow (1.24)^z \qquad (9)$$

indicating an increase of the shear rate and of the shear stress from level 0 to level 8 by the factor 5.6, a value which is reasonably well in agreement with experimental data.

The Reynolds number is proportional to

$$Re \hookrightarrow D_z^{-1} n_z^{-1} \qquad (10)$$

and after insertion of numerical values:

$$Re \hookrightarrow (0.175)^z \qquad (11)$$

Thus the model predicts a decrease of the Reynolds number by the factor 10^{-6} from the aorta to the capillary level. This is in agreement with values given for the dog by Whitmore (1968). Sources of data for the morphological properties of the canine arterial system are found in Patel et al.(1963). For human beings data are given by Patel et al. (1964), Iberall (1967) and Westerhof (1969). A survey of literature is found in McDonald (1974).

The number of levels in the arterial tree deserves some more discussion. An interesting and illustrative comparison is possible for the pulmonary arterial tree. Vadot (1967) counts 9 levels similar to the systemic arterial tree. Lefevre (1982) mentions 20 levels of branching points which are mainly dichotomous, an assumption which was earlier proposed and discussed by Pollak et al.(1968). On one hand it seems surprising that a morphologic distinction between 9 and 20 levels seems to be difficult to decide. On the other hand it is easy to understand that the final overall effect in terms of the number of branches of 8 quadruple branching points is equal to 16 dichotomic branching points.

As far as the systemic arterial tree is concerned, a classification by McDonald (1974) assumes 9 levels similar to Vadot (1967). Iberall (1967) introduced a model with 11 levels.

The following statement by McDonald (1974) is certainly quite correct: " The large numbers of X-ray arteriographs that are made annually in all major medical centres would suggest a large source of data which has not, to my know-ledge, been systematically organised."

1.4 The branching ratio

In bifurcations the total cross section area of both daughter vessels in most examples is somewhat larger than the cross section area of the parent vessel. The socalled branching ratio

$$d = 2r_1^2/r_0^2 \tag{1}$$

is quoted to have values around 1.1 to 1.2 at the aortic bi-furcation (Wetterer and Kenner 1968) and values close to 1.26 at many major arterial bifurcations (McDonald 1974).

This particular value 1.26 of the branching ratio has some interesting implications. It can be shown that for a

given volume, resistance as well as energy dissipation are
minimized if the branching ratio of a bifurcation corresponds
to the value 1.26. Rosen (1967) showed that the minimization
of the following cost functional F can be used for this calculation:

$$F = q^2 R + K(l_o r_o^2 \pi + l_1 r_1^2 \pi) \tag{2}$$

where R is the resistance of the vessel, l_o, r_o and l_1, r_1 are
length and radius of parent and daughter vessel. Using cal-
culus of variation one obtains for the case of symmetric
branching the following condition for the minimization of F:

$$r_o^3 = 2r_1^3 \tag{3}$$

or

$$r_o = 1.26 \, r_1 \tag{4}$$

or

$$d = 2r_1^2 / r_o^2 = 1.26 \tag{5}$$

Thus , the branching ratio 1.26 corresponds to the optimal
value.

  Gessner (1981) recently pointed out that the condition
of equal wall shear stress in all vessels leads to the same
result. Since

$$\tau_{so} = -4 \, \eta \, q / (r_o^3 \pi) \tag{6}$$

as shown in section 1.2 (eq.7), is the shear stress in the
parent vessel, the shear stress in each daughter vessel is

$$\tau_{s1} = -2 \, \eta \, q / (r_o^3 \pi) \tag{7}$$

if both are related to the total flow q through the branches.
If both shear stresses are assumed to be equal, we again

obtain

$$r_o{}^3 = 2r_1{}^3 \tag{8}$$

In the case of two unequal daughter vessels both conditions, minimization of resistance and equating of wall shear stress lead to the more general condition

$$r_o{}^3 = r_1{}^3 + r_2{}^3 \tag{9}$$

We had observed in section 1.3 that, actually in vivo, a slight increase of the shear stress from the aorta towards the smaller vessels can be observed. However, this increase is not very marked.

There are some additional conditions and aspects concerning branching and bifurcations, which are worthwhile to be mentioned. Interesting is the behaviour of the characteristic impedance (for definition see section 2.3 and 3.2) and the influence of wave reflections (see section 3.5). Wetterer and Kenner (1968) could show, that - assuming equal geometry and elastic properties of parent and daughter vessels - the condition

$$d = \sqrt[5]{2} = 1.15 \tag{10}$$

guarantees reflection free condition at a bifurcation.

2. THE WALL OF SMALL ARTERIES

2.1 Introduction

The physical behaviour of arteries can be observed in vivo or examined in vitro under the following conditions: Vessels can be freely extensible like mesenteric arteries or coiled uterine arteries. In this ease it will be observed that during pressure pulsations each pulse pressure increase will lead to a lengthening of the freely distensible segment. As the other extreme, vessels may be longitudinally constrained. Here length is constant. Since arteries have the common property that increasing pressure distends them both in longitudinal and in circumferential direction we will observe that the force in length direction which is related to the constraint , increases as the pressure decreases. If an arterial segment is cut out of a tissue it retracts by 30 to 40% and in some cases even more. During transition of an arterial pressure pulse the longitudinal wall stress may either show pulsatile decrease or increase, depending on the mean arterial pressure and on the contractile condition of the arterial smooth muscles.

2.2 Arterial wall structure

The inner surface of any artery is lined with endothelial cells which usually tend to be arranged with their large ahis in the direction of the blood stream. An example of endothelial cell lining in a small artery is schematically shown in fig.5.

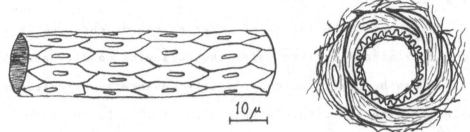

Fig.5 Endothelial lining of a small artery (left),
layers of a small artery (right)

The inner layer of the artery including the endothelial cells
and a small subendothelial layer is called the tunica intima.
This subendothelial layer contains in most small arteries a
more or less expressed internal elastic membrane which appears
folded in the relaxed state (as arteries usually are fixed
undistended for preparation for microscopic slides). The next
large layer - the tunica media - is the main load-carrying
layer of small arteries. It contains layers of elastic fiber
nets and smooth muscles which appear to interconnect these
elastic fibers. One of the best descriptions is found in
Benninghoff (1930). Besides the elastic fibers and smooth
muscles the tunica media also contains collagen fibers. We
can distinguish elastic and muscular arteries. In general
small arteries belong to the group of muscular arteries. As
the percentage of smooth muscles increases from larger arteries towards smaller arteries and arterioles the content
of elastic tissue decreases from about 45% to 20% according to Noordergraaf (1978). The content of collagen fibers

decreases from about 30% in the aorta to about 20% or somewhat less in the arterioles. The smooth muscle cells increase in total mass from about 25% in the aorta to around 60%
in the arterioles.

The media is surrounded by the socalled tunica adventitia (or simply adventitia) which extends into the surrounding
tissue. It seems important for experimental measurements
and for the calculation of tensions and stresses, that the
determination of a clear limit between the vessel wall and
the surrounding tissue is, quite often, rather difficult.

The overall relation between pressure and radius of a
small artery can be characterized under quasistatic conditions as being S-shaped as shown in fig.6 which shows the behaviour of a small rat carotid artery (Weizsäcker and Pascale
1982).

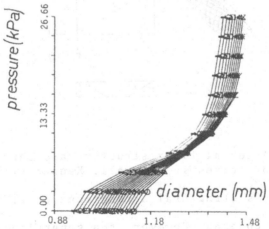

Fig.6 Pressure-diameter curves from a rat carotid
       artery (Weizsäcker and Pascale 1982)

The artery was examined under longitudinal constraint with

different prestretch in longitudinal direction. The length
of the segment was adjusted between 1.4 and 2,6 cm. The larger
the longitudinal stretch the smaller the diameter.

    The S-shape of the curves can be explained by a three
phase process. The tissue components show following sequence
of extension: flattening and stretching of elastic fibers,
stretching of smooth muscles. Finally, the extension is limit
ted by the extension and stretching of collagen fibers.

2.3 Mechanical behaviour of arterial walls

    Most structures of the arterial wall are arranged in
spiral loops as indicated in fig. 7 (Kenner 1967).

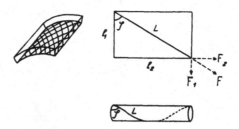

Fig.7 Arterial fiber structure and paralellogram
          of forces in the wall. Kenner (1967)

The details of this structural arrangement of fibers may
differ from area to area. However, the general outline seems
to be quite similar (Benninghoff 1930, Kenner 1967, Rhodin
1980, Hudetz and Monos 1982, Weizsäcker and Bescale 1977,1982

The forces in the vessel wall can be described by the following equations if the wall thickness (h) is small compared to the diameter (2r).

The circumferential force carried by a segment of length l is

$$F_1 = r\ l\ p \qquad\qquad (1)$$

the corresponding circumferential stress per unit cross section area is

$$\sigma_1 = p\ r/h \qquad\qquad (2)$$

where r is the radius and h the wall thickness. Since, in the relaxed condition, $r/h \triangleq 10$ the circumferential wall stress has about 10 times the magnitude of the arterial pressure.

In the longitudinal direction the total force F_2 is composed of two components

$$F_2 = r^2 \pi\ p + F_{2ext} \qquad\qquad (3)$$

where F_{2ext} is the force which is related to, or generated by, the longitudinal constraint of the vessel. As has already been mentioned in the introduction the effect of this force can be demonstrated if a vessel is cut out from the tissue. Then it retracts by 30 to 40% or even more depending on the contraction state of the smooth muscles.

Per unit cross section area the corresponding stresses carried by the material are:

$$\sigma_2 = p\ r/(2h) + F_{2ext}/(2r\,\pi\,h) \qquad\qquad (4)$$

The spiral structure of the vessel walls leads to an interesting observation concerning the passive behaviour (van
Loon et al. 1977,Weizsäcker and Pascale 1977,1982),which is
shown in fig. 8.In a rat carotid artery the length-force
relation of the vessel was recorded. It depends on the internal pressure in such a way that below a certain length the
increasing pressure tends to decrease the longitudinal force
F_{2ext} whereas above this particular length the effect is reversed. We call this unique point in the diagram "the crossing
point".

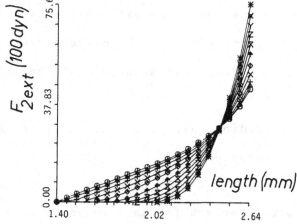

Fig. 8   Longitudinal force F_{2ext} versus length,
         pressure range 0 to 200 mm Hg, pressure
         steps 20 mm Hg.Weizsäcker and Pascale,1982

In the rat the behaviour can be observed in different vessels
as examined currently in our department. Van Loon et al.(1977)
have observed the same phenomenon in dog arteries. An explanation is given in the following sketch, fig. 9 (Kenner 1967).
As a vessel is distended by internal pressure, the fibers tend
to adjust themselves into the direction of the parallelogram

of forces in the wall. The equilibrium condition will be determined below (eq.9). If an artery is extended over the
equilibrium length,the increasing internal pressure will
tend to "contract" the artery. If the length of the artery is
below the equilibrium point then increasing internal pressure
will extend or relax the artery as explained in fig.9.

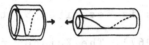

Fig. 9 Scheme, explaining effect of pressure
on length-forces in an arterial wall

The following simplified calculation estimates the
average angle of the spiral fibers in the equilibrium state
if these fibers are assumed as unrestrained by elastic forces. Under the condition of free extension of an unrestrained
vessel segment the external length force F_{2ext} is assumed
as zero:

$$F_{2ext} = 0 \tag{5}$$

then from eq. 3 and 4 follows

$$\sigma_2 / \sigma_1 = 1/2 \tag{6}$$

Using the relations shown in fig. 7 we can, furthermore,
derive

$$(7) \qquad \sigma_1 = F_1/hl_2 = F \cos\varphi /(hl \sin\varphi)$$

and

$$(8) \qquad \sigma_2 = F_2/hl_1 = F \sin\varphi /(hl \cos\varphi)$$

Therefore, the equilibrium condition is

$$\sigma_2/\sigma_1 = tg^2 \varphi \qquad\qquad (9)$$

and with the condition of free extension (no constraint)
$\sigma_2/\sigma_1 = 1/2$ we find $\varphi = 35.2^{\circ}$, a value which is reasonably
close to measured values (Kenner 1967, Kenner et al. 1982,
Hudetz and Monos 1982).

The interaction between forces and extensions in three
orthogonal directions is described by the Poisson ratio (see
Fung 1981, Kenner 1967). The Poisson ratios of the arterial
wall tissue show anisotropic behaviour and depend on the
extension. At small values of the extension, nearly no interaction between orthogonal directions can be observed. The
Poisson numbers under this condition may be close to zero.
During extension the Poisson numbers increase. In an extended
spiral fiber structure like that shown in fig. 7 Poisson
ratios of larger than 1 can be observed (Kenner 1967).

Due to the elastic distensibility of the spiral fiber
structure, a freely extensible arterial segment which is
continuously filled and extended by fluid will stretch in
longitudinal as well as in circumferential direction. In
contrast, a rubber tube will, under the same conditions extend only in circumferential direction because, in this case,
the material is homogeneous   The interaction between orthogonal directions is described by the value 0.5 of the Poisson
number in all directions (Wetterer and Kenner 1968).

Fig. 10 shows schematically the relation between pressure and external longitudinal force F_{2ext} in a rat carotid
artery under three different conditions. Below the crossing
point - explained in fig.8 - an increase of the distending
pressure decreases the external constraining force F_{2ext}.

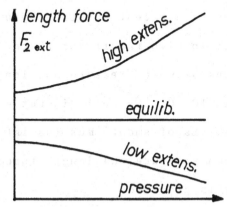

Fig. 10 Longitudinal force as a function of the
distending pressure . Parameter: length

Exactly at the crossing point, the external force is independent of pressure. If the vessel segment is extended beyond
the crossing point, an increase of the internal pressure will
further increase the external constraining force as was
schematically explained in fig. 9. We assume that the
crossing point is close to the natural length of the vessel.
It seems possible, however, that the contractile status of
the vascular smooth muscles may influence the position of the
crossing point.

It can be expected that the three possibilities of
longitudinal force variation to changes in arterial pressure
may also be observed during pressure pulse transients.

Wetterer and Kenner have already in 1968 critizised naive reports - usually badly supported by experiments - which from
time to time mean to "prove" the existence of fast contractions of vascular smooth muscles which follow the pulse wave
and are interpreted as kind of peristaltic movement. These
reports, including a recent one on rhythmic contractile
activity of the in vivo rabbit aorta, can be explained by
the interactions between pressure and length force as demonstrated in fig. 10. Of course it is not surprising that activation or paralysis of smooth muscles influence the phase
relation between pressure and length force if a shift of the
crossing point is produced.

2.4 Relations between vessel wall properties, flow and wave
     velocity - some "rule of the thumb - equations"

     Although in small vessels the effect caused by the
viscosity of the fluid and of the wall material (both of
which will be discussed later) is rather marked, some simple
relations should be mentioned here. The relations which are
valid only for ideally elastic frictionless tubes, neverthe+
less, appear useful to explain some fundamental phenomena
and to check orders of magnitude (Wetterer and Kenner 1968).

     The relation between pressure amplitude Δp and flow
amplitude Δq in an ideally elastic tube without reflections
is defined as characteristic impedance:

$$Z = \Delta p / \Delta q \qquad (1)$$

Provided viscosity can be neglected, the characteristic impedance is related to the wave velocity c in the tube by

$$Z = c\rho / A \qquad (2)$$

where $A = r^2\pi$ is the cross section area of the vessel. The properties of the vessel which is filled with fluid of density ρ can be described by the compliance C per unit length

$$C = \Delta A / \Delta p \qquad (3)$$

and by the socalled effective mass per unit length

$$M = \rho / A \qquad (4)$$

The wave velocity of an ideally elastic tube c is related to these magnitudes by

$$c^2 = 1 / MC \qquad (5)$$

We define the velocity amplitude of a flow pulse as the peak amplitude of the velocity averaged over the cross section area of the tube:

$$\Delta v = \Delta q / A \qquad (6)$$

The equations described above yield the following relation between pulsatile variations of the cross section area (ΔA), the velocity pulse amplitude (Δv), and the wave velocity c (Wetterer and Kenner 1968):

$$\Delta v / c = \Delta A / A = 2 \Delta r / r \qquad (7)$$

If, in a pulsating artery, pressure and diameter is simultaneously recorded we observe a relation between the two variables which indicates by its gross steepness the elastic distensibility of the tube. Usually the record shows a hys-

teresis loop in which pressure leads diameter. As will be
discussed below the area of the loop is   proportional to
the viscous energy loss produced by viscous friction within
the wall material. In elastic arteries the loop usually has
a smaller opening than in small muscular arteries. An example taken from Wetterer et al. (1977) is shown in fig.11.

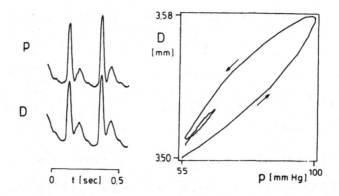

Fig. 11 Pressure (p) and external diameter (D) of
a common carotid artery of a dog. Wetterer
et al. (1977)

In this example we find that the relative radius pulsation is
in the order of magnitude of 2% in the femoral artery of a
dog, corresponding to an area pulsation of 4%. This is in
agreement with eq. 7 if the ratio equals

$$\Delta v/c = 80(cm/s)/2000(cm/s) \tag{8}$$

These values agree reasonably well with actual measurements
in the dog, although the wave velocity is rather high.(See
also section 3.6). We have to mention here that the proper-

ties of the aorta-iliaca-femoral tube change in several ways

from the aortic root towards the periphery (Wetterer and

Kenner 1968, Kenner 1979). As shown schematically in fig.12

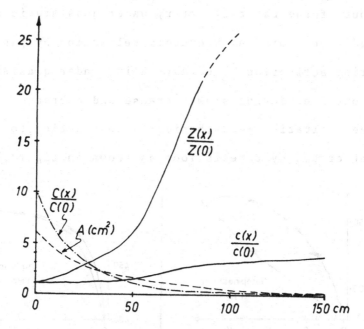

Fig. 12 Properties of the aorta-iliaca-femoral
         tube; geometric and elastic tapering

the compliance C, and the cross section area A decrease,

whereas the pulse wave velocity c and the characteristic im-

pedance Z increase towards the periphery (as estimated for

a human being). The tapering effect explains the high wave

velocity in smaller muscular arteries.

2.5 Dynamic properties of small arteries

     There is an abundant literature about the dynamic elas-

tic properties of blood vessels. Recent reviews include Cox,

Monos and Kovach,Hudetz and Monos, Busse et al.,Newman and

Greenwald, in Kenner et al.(1982); furthermore Dobrin 1978

and Basar 1981.

Busse et al.(1982) have recently reexamined the dynamic

behaviour of the rat tail artery under quasistatic and dyna-

mic condition, both under extreme relaxation by papaverin

and during activation by noradrenalin. Under quasistatic

conditions,i.e. during slow increase and decrease of pres-

sure,the activation leads,besides a contraction,to a marked

widening of the hysteresis loop as shown in fig 13,left.

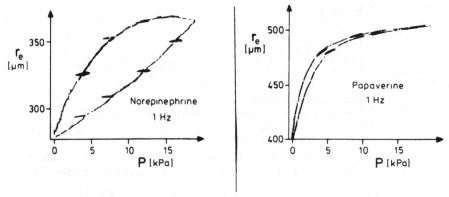

Fig. 13   Quasistatic pressure-radius curves of
          the rat tail artery (Busse et al.1982)

The superimposed small dynamic loops, recorded at 1 Hz did

not change their slope very markedly. A similar observation

was also reported by Newman and Greenwald (1982).This obser-

vation implies that there are two different mechanisms which

have to be considered in order to explain the viscoelastic

behaviour of small vessels with and without activation of

smooth muscle contraction. Busse et al.(1982) assume that

the widening of the hysteresis loop during muscle contracti-

on is due to the time- and load-dependent reattachment of
cross bridges of contractile elements, actin and myosin in
the smooth muscles. Therefore the pattern of the quasistatic
hysteresis loop can be attributed to the properties of the
contractile element (CE) of the smooth muscles. The quasistatic elastic modulus, visually expressed in fig.13 by the
steepness of the radius-pressure relation is markedly lower
during contraction than in the relaxed state.

The steepness and, therefore, the dynamic elastic moduli during the small 1 Hz cycles seems to depend mainly on the
mean pressure and is not much influenced by the contraction
of the vessels. The behaviour of these small dynamic loops
can be ascribed to the properties of some passive elastic
elements which can be modeled as between in series to the contractile element.

2.6 Linear modeling of the dynamic elastic behaviour

As a first approximation of the description we assume
piecewise linearity and consider only small extensions

$$\epsilon = dl/l \tag{1}$$

Between ϵ and the stress σ and the elastic properties we
can observe the following general relation:

$$\sigma = \epsilon E_d + \eta_w \, d\epsilon/dt \tag{2}$$

Complex Fourier transformation and division by ϵ leads to

$$E = E_d + j\omega\eta_w \tag{3}$$

E is the complex modulus of elasticity. E_d is the dynamic
modulus of elasticity. η_w is the coefficient of wall viscosity. The equation implies the application of a model as
shown in fig. 14 which is usually called a Voigt solid(Fung
1981). If the magnitudes E_d and η_w are allowed to be func-

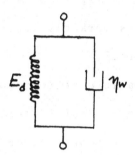

tions of the frequency ω then any
viscoelastic model can be fitted
into the above equation. This means
that both components of the model
in fig.14 are assumed as frequency
dependent.

Fig.14

In my opinion a certain confusion may possibly be caused by the
sometimes indiscriminate exchange of the viscous (dashpot)
and the contractile element (CE) in models given the same
names. Fig. 15 shows, for comparison, Voigt and Maxwell models from Noordergraaf (1978) left and from a paper by Cox
(1982) right.

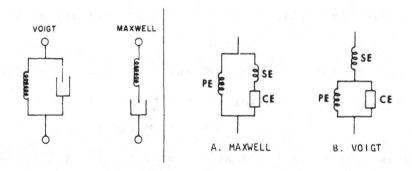

Fig.15 Models with and without contractile element

Certainly, whenever the use is properly defined, the difference is not a mistake but, nevertheless, confusing. A possible confusion may be even increased because the actual Maxwell element is a model of an elastoviscous fluid. It seems
to me obvious that in a model of a smooth muscle the series
elastic element (SE in fig.15) which describes the elastic
behaviour of the dynamic 1 Hz cycles in fig. 13 ought to be
a viscoelastic element. The parallel elastic element (PE)
describes the properties of the muscle during relaxation.
From the description of Busse et al.(1982) it appears that
PE as well as CE exhibits complex properties during contraction.

In terms of the model of fig.15 Busse et al.(1982) describe the following results in a small artery: E_d and η_w
increase with increasing wall stress. At a given wall stress,
E_d is virtually independent of the frequency while η_w decreases markedly with increasing frequency. This behaviour of η_w
is called thixotropy or pseudoelasticity and agrees basical-
ly with Bergels wellknown observations (1961). Smooth muscle
activation does not markedly change either E_d or η_w in the
small muscular arteries.

2.7 A note on incremental moduli

According to a report by O.Frank (1906), W.Roentgen
(1876) - the discoverer of the X-rays - was the first to reco-
gnize the problems related to the description of highly ex-

tensible materials. O.Frank himself was the first to introduce an incremental modulus of elasticity. As discussed by
Kenner (1967) and by Wetterer and Kenner (1968), there are
two possibilities to define Franks incremental moduli. The
difference apparently, was not recognized by Frank (1906,
1920).

Since, for certain problems, the difference plays a
role, we discuss the problem in the following.

The two definitions of incremental moduli are the
"force related modulus"

$$E = (dF/A)(1/dl) \tag{1}$$

and the "stress related modulus"

$$\mathcal{E} = d\sigma \, dl/l \tag{2}$$

with F force, A cross section area of the material, $\sigma = F/A$
stress. The interrelation between the two moduli in the case
of one dimensional stretch is given by the equation

$$d\sigma = dF/A - \sigma \, dA/A \tag{3}$$

and, therefore, valid for extensions in one direction and
with the assumption that the material is incompressible:

$$\mathcal{E} = E + \sigma \tag{4}$$

The problem which has been solved in a paper by Kenner
(1967) is the fact that the relation of the two modulus
systems in two dimensional extensions (of thin-walled incompressible arteries) can only be described by systems of matrix equations. Both systems of matrix equations can be re-

lated to the stress energy function of the material.

For a longitudinally constrained vessel the following simplified equations describe the circumferential extension (index 1) in terms of the force related modulus system (length direction index 2, μ_{ij} are the Poisson numbers which are dependent on the direction and differ slightly between both modulus systems):

$$dr/r = dF_1/(A_1E_1) - \mu_{12} dF_2/(A_2E_2) \qquad (5a)$$

$$dl/l = 0 = -\mu_{21}dF_1/(A_1E_1) + dF_2/(A_2E_2) \qquad (5b)$$

Further insertion yields

$$dr/r = [dF_1/(A_1E_1)](1 - \mu_{12}\mu_{21}) \qquad (6)$$

and

$$dF_1/A_1 = d(lrp)/hl = (r\ dp + p\ dr)/h \qquad (7)$$

Insertion of these equations leads finally to the definition of the circumferential incremental modulus in a length constrained tube (in the force related definition):

$$E_1 = \frac{r}{h}(r\ dp/dr + p)/(1 - \mu_{12}\mu_{21}) \qquad (8)$$

The corresponding equation for the stress related modulus is

$$\mathcal{E}_1 = \frac{r}{h}(r\ dp/dr + 2p)/(1 - \mu_{12}\mu_{21}) \qquad (9)$$

According to Wetterer and Kenner (1968) two facts are noteworthy: 1) The usual definition of the incremental modulus E_{inc} which is also valid for thin walled tubes, corresponds to the force related modulus E_1 (e.g. Bergel 1961, Monos and Kovach 1982). 2) The usual omission of the factor p (pressure) which comes into the equation from the correct differentiation of the circumferential force (or stress - in the stress related modulus system) may produce an error of 10 to 20% in the calculated modulus.

Busse et al. (1982) have used the force related incremental modulus to describe the dynamic properties of small vessels taking into account finite wall thickness (h). r_i and r_e are internal and external wall radius, respectively. μ is the Poisson number and φ is the phase angle between sinusoidal pressure and radius variation. The correct definition of the incremental dynamic modulus according to Busse et al. (1982) then is as follows: The real part

$$E_d(\omega) = ((2r_e^2 r_i)/h(r_i^2+r_e^2))(1-\mu^2)(p+r_i(\Delta p/\Delta r_i)\cos\varphi \qquad (10)$$

and the imaginary part of the modulus

$$\omega \eta_w = ((2r_e^2 r_i)/h(r_i^2+r_e^2))(1-\mu^2)r_i(\Delta p/\Delta r_i)\sin\varphi \qquad (11)$$

The assumption that the Poisson number is invariant, independent of stretch, isotropic and equal 0.5 in all directions is certainly not correct in vessel walls, and introduces another error - which, by the way, is quite frequently found in the literature. However, the experimental determination of Poisson numbers is rather difficult.

2.8 Critical closing pressure

Most small muscular arteries and arterioles have the capability of complete closure by contraction. Wetterer and Kenner (1968) have extensively discussed the concept of a critical closing pressure. Burton (1951) had assumed that

during a decrease of the internal pressure or during an increase of active wall tension the radius may become smaller
and smaller until at a certain moment "instability" is reached as the radius shrinks to zero. On the one hand we actually can observe a zero-flow pressure intercept. On the
other hand the original concept by Burton (1951,1962) is mis-
leading because of the somewhat abstract and ideal assumption
that a vessel may in the contracted state have a zero radius. Furthermore, the use of a "Laplace" tension defined as a
force per unit length does not take into account that the
vessel wall material has a finite cross section. Wetterer and
Kenner (1968) have critizised the reference to "Laplaces law"
which may be appropriate for the description of soap bubbles
but not for the analysis of blood vessels with finite wall
thickness. It seems particularly important that the mean radius of the tunica media never can become zero even in the
maximally contracted state.

Meanwhile similar critique has also come from other authors as summarized by Gow (1980). In conclusion we can assume that a closure of an artery is possible without any "instability". The endothelium then is found to be extremely folded.
An opening is possible at the same pressure as the closing
pressure because at some point upstream the lumen of the contracted vessel may extend conically into the next larger open
segment. Large arteries do not have the capability of complete closure.

Burton (1951) in any case has the merit to have initiated a discussion on a very important topic. Closure of vessels and its relation to the properties of the vessel wall
plays a very important role under pathologic conditions as e.
g. the development of hypertension.

It should be added here that the closure of small vessels can be brought about by three processes: 1) the first
process is the more or less concentric closure by the contraction of the media. 2) In striated muscles and particularly in the myocardium the tissue pressure generated by the
muscle contraction leads to a collapse of small vessels. In
other words, under these conditions small arteries have to
be considered as collapsible vessels. The two processes are
schematically shown in fig.16 . 3) In some small vessels
swelling of endothelial cells has been observed.

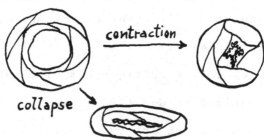

Fig. 16 Constriction and collapse of a small vessel

2.9 Pressure-flow relation at the peripheral resistance
Many measurements of the peripheral pressure-flow relation have been made in local arterial beds and in the
whole arterial system as summarized by Wetterer and Kenner

1968,McDonald 1974,Kenner 1979, Kenner et al.1982. One exam-

ple of a typical quasistatic measurement in the lung circula-

tion by Wezler and Sinn 1953 is shown in fig.17.

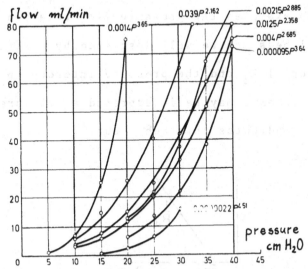

Fig.17 Pressure-flow relations (Wezler a.Sinn 1953)

The experiments were made in the plasma perfused rabbit lung.

Besides each curve from single measurements the equation of

the parabola is written which is used to approximate the func-

tion. The shape of the curves can be explained by the passive

elastic distensibility of the small arteries and arterioles.

Green et al.(1944) had first proposed a description of

the pressure-flow relation in the form of

$$q = a \, p^n \tag{1}$$

As discussed by Ronniger (1955) the following relations can

be derived from this equation: the absolute value of the re-

sistance $\qquad 1/R_{abs} = q/p = a \, p^{n-1}$ $\qquad$ (2)

and the differential value of the resistance

$$1/R_d = dq/dp = a \, n \, p^{n-1} \tag{3}$$

and the relation between the two

$$R_d/R_{abs} = 1/n \qquad (4)$$

With respect to the results described below it seems

that in many examples the pressure-flow relation of periphe-

ral arterial beds can be well described by a straight line

with the slope $1/R_d$ and the pressure intercept p_c. With some

caution the index c may be interpreted as closure. The follo-

wing fig. 18 indicates this behaviour.

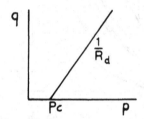

Fig.18   Pressure-flow relation

The function can be described by the following equation:

$$q = (p - p_c)/R_d \qquad (5)$$

For the measurement in the whole circulation, Wetterer

and Pieper (1953) have developed a method which has recently

been revived and extensively used by Dujardin (1982). The

principle of the method is based on the indirect measurement

of diastolic outflow from the arterial system during the ar-

tificial introduction of slow volume variations with a pump

connected to a large artery. The frequency of the pump is

in the order of 0.2 Hz. Since the variations of pressure and

volume are rather slow, a simple windkessel model is used for

the evaluation (Wetterer and Kenner 1968, Dujardin 1982) as shown in fig. 19.

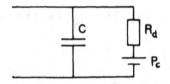

Fig. 19 Windkessel model

In this model the zero pressure intercept is simulated by

a battery. In the notation of Dujardin (1982) the non linear

compliance of the windkessel is

$$C = C_o (1 - k\ p) \qquad (6)$$

where k is a constant and p is the pressure. The total flow

into the windkessel is then described by the equation

$$q = (p - p_c)/R_d + C\ dp/dt \qquad (7)$$

During each diastole the input into the system equals the

pump flow q_P . The first term on the right side of the wind-

kessel equation is always the outflow q_R through the periphe-

ral resistance. With respect to the second term we introdu-

ce the following approximation and abbreviation:

$$dp/dt \ \hat{=} \ \Delta p/D \ = \ S \qquad (8)$$

where D is the duration of the diastole and Δp is the pres-

sure difference during diastole. Then the windkessel equati-

on reads during diastole as follows:

$$q_P = q_R + C\ S \qquad (9)$$

Fig. 20 shows an original recording from Wetterer and Pieper

(1953) in a dog. At 90° the maximum volume has been pumped

into the aorta; 270° indicates the maximum volume in the pump.

In the original method Wetterer and Pieper (1953) used a pair

of pulses with the same mean diastolic pressure (pulse A and

B).For both pulses the outflow q_R is the same.Furthermore, if

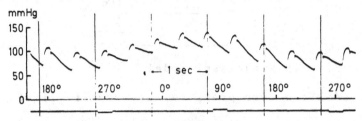

Fig. 20 Blood pressure during sinusoidal pumping
(Wetterer and Pieper 1953)

pulse A is recorded during positive pump flow, pulse B corresponds to the negative pump flow of the same magnitude.The
windkessel equation then yields

$$C = (q_{PA} - q_R)/S_A \quad \text{and} \quad C = (q_{PB} - q_R)/S_B \qquad (10)$$

where the indices A and B relate to the two pulses. From the
equations (eq.10) we find two solutions.One permits to calculate the total compliance C of the arterial system.The other
solution yields the total peripheral flow q_R during diastole.

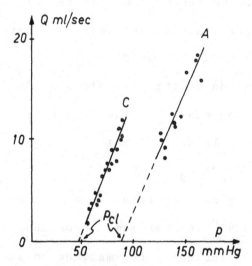

Fig.21 Pressure-flow relation (Wetterer a.Pieper 1955)

Fig. 21 shows an example of the relation between pressure

and flow in an anesthetized dog before and after administra-

tion of adrenalin (A) as measured by Wetterer and Pieper

(1953,1955). It is interesting to note that the pressure-

flow relation actually is found to be straight and that vaso-

constriction leads to a shift of the whole relation more or

less parallel. This indicated that R_d may, during vasocon-

striction, stay constant. Dujardin (1982) used a parameter

estimation method for the evaluation of the windkessel equa-

tion. Fig.22 shows the pressure-flow relation measured in

normal rats (WKY,circles) and in spontaneously hypertensive

rats (SHR,squares). In this example the slope of the pressure

flow relation is different for the two conditions indicating

an increase of the differential value of the resistance R_d

in spontaneous hypertensive rats compared to normal rats.

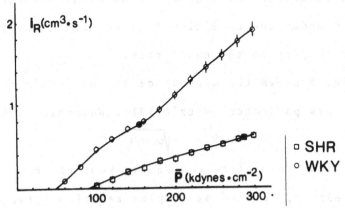

Fig. 22 Pressure-flow relation in rats(Dujardin
1982)

The method described in this section allows clearly to

characterize the overall behaviour of the small arteries.

3. FREQUENCY DYNAMICS OF THE PERIPHERAL CIRCULATION

3.1 Introduction

With some exceptions we have so far restricted our discussion to quasistatic phenomena. However, the mechanics of
the normal circulation is characterized by pulsatile activity
of the heart. It is important to note that a characteristic
dimensional relation exists between heart rate and the proportions of the arterial system (Kenner,1979). The normal
resting heart rate of animals is reciprocally related to the
length dimension of the body. Therefore, heart period, time
constants of diastolic arterial pressure decay and autooscil-
lation periods of the arterial pressure are proportional to
each other and to the cube root of the body mass. Therefore,
the contour of the arterial pulses is similar in all animals.
Likewise impedance and pressure transmission functions are
similar - under the condition that the frequency is normalized with respect to the heart rate.

Fig.23 shows the dependence of the aortic value of the
unsteadiness parameter (Witzig 1914, Womersley 1957) α :

$$\alpha = r\sqrt{\omega\varrho/\eta} \qquad (1)$$

which is proportional to the square root of the aortic radius.

Impedance, defined as complex relation between pressure
and flow, and pressure transmission function,defined as complex relation between two pressures, are well established
functions used to describe the properties of the arterial

system. Both functions depend on two basic components: the characteristic impedance and the peripheral resistance.

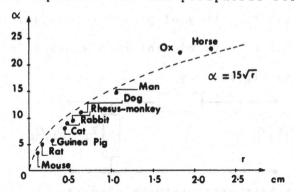

Fig.23 Parameter α as function of the radius of the aorta (Kenner 1972)

In discussing frequency dynamics we will have to cover 1) the frequency region of the heart rate and its harmonics and 2) the low frequency region. - In our discussions so far the small arteries, the arterioles and the microcirculation have been assumed as time invariant localized elements which determine the magnitude of the peripheral resistance. This part of the article will extend the view and include reactions and responses of the small arteries and resistance vessels which are mediated by nerves or by autoregulatory mechanisms.

Summaries of frequency dynamics of the peripheral circulation are given by Attinger (1973) and Basar (1981).

3.2 High frequency behaviour of a transmission line

In spite of the possibilities to calculate flow and pressure in rather complex tube systems using increasingly

detailed and accurate models and methods the fundamental analysis of a 4-terminal network transmission line still is
highly useful for the study of pressure and flow in arteries.

A 4-terminal network element as discussed in the following is shown in fig.24.

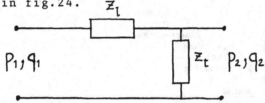

Fig. 24 A 4-terminal network element

Indices 1 and 2 refer to input and output, respectively.The
relation between pressure and flow at input and output can
be described by the matrix equation

$$\begin{pmatrix} p_1 \\ q_1 \end{pmatrix} = \begin{pmatrix} p_2 \\ q_2 \end{pmatrix} A \tag{1}$$

where

$$A = \begin{pmatrix} 1 + z_1/z_t & z_1 \\ 1/z_t & 1 \end{pmatrix} \tag{2}$$

If the 4-terminal element has the infinitesimal length dx
then the transfer matrix A for one such element is

$$A = \begin{pmatrix} 1 & z_1 dx \\ dx/z_t & 1 \end{pmatrix} \tag{3}$$

By a matrix operation it is possible to calculate the matrix
A for a chain of such infinitesimal elements which, put together in series, constitutes a transmission line of length L.

$$A = \begin{pmatrix} \cosh \gamma L & Z \sinh \gamma L \\ (1/Z)\sinh \gamma L & \cosh \gamma L \end{pmatrix} \tag{4}$$

where Z is the characteristic impedance:

$$Z = \sqrt{z_1 z_t} \quad . \tag{5}$$

The complex transmission coefficient:

$$\gamma = \sqrt{z_1/z_t} = \beta + jk \tag{6}$$

The two components of the transmission coefficient are the real part β, the damping coefficient, and the imaginary part k, the phase coefficient. The latter is related to the wave velocity c by

$$c = \omega/k \tag{7}$$

The role of the transmission coefficient is best described by the relation between complex pressure P_1 at the entrance and P_2 at the end of a tube with a matching impedance Z at its end:

$$P_2 = P_1 e^{-\gamma L} \tag{8}$$

In general the pressure-flow relation at the end of a tube can be described by a (real or complex) impedance R:

$$Q_2 = P_2/R \tag{9}$$

and after multiplication of the matrix equation we find the complex pressure transmission function:

$$P_1/P_2 = \cosh \gamma L + (Z/R)\sinh \gamma L \tag{10}$$

Furthermore, the flow transmission function:

$$Q_1/Q_2 = \cosh \gamma L + (R/Z)\sinh \gamma L \tag{11}$$

and the input impedance

$$P_1/Q_1 = Z \frac{\cosh \gamma L + (Z/R)\sinh \gamma L}{\sinh \gamma L + (Z/R)\cosh \gamma L} \tag{12}$$

In order to apply these equations we have to describe the longitudinal impedance z_1 and the transverse (circumferential) impedance z_t. Fig.25 shows a generalized 4-terminal element of a viscoelastic leaking tube.

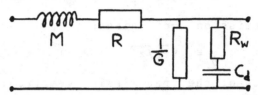

Fig. 25 4-terminal element of a viscoelastic
          leaking tube

The longitudinal impedance is composed of resistance to flow
R and effective mass M per unit length

$$z_1 = R + j\omega M \qquad (13)$$

As rough approximation one possible assumption is

$$R = 8\eta/r^4\pi \qquad (14)$$

The effective mass per unit length is defined as

$$M = \rho/A \qquad (15)$$

where $A = r^2\pi$ and ρ is the density of blood. By introducing
the unsteadiness parameter α as defined in section 3.1 :

$$z_1 = j\omega M(1 - j \, 8/\alpha^2) \qquad (16)$$

For the transverse impedance we consider viscoelastic
distensibility of the vessel wall. The compliance C is

$$C = 1/(1/C_d + j\omega R_w) \qquad (17)$$

C_d is the dynamic compliance and R_w is the internal frictional resistance of the vessel wall. - The electric analog for
a parallel spring and dashpot is a capacitor and a resistor
in series as shown in fig.25.

In addition, leakage by outflow from the tube through
small side branches is taken into account by the outflow ad-
mittance G. Altogether we can write:

$$z_t = 1/(G + j\omega C_d/(1 + jtg\varphi)) \qquad (18)$$

The tg of the phase angle between pressure and radius variation due to the viscoelastic wall properties equals

$$tg\,\varphi = \omega\eta_w/E_d = \omega R_w C_d \qquad (19)$$

Under the condition that the resistance to flow R is small and can be neglected, and that tg φ as well as G are small, we find the characteristic impedance for larger vessels

$$Z \stackrel{\wedge}{=} Z_o(1 + jtg\varphi/2 + jG/(2\omega C_d)) \qquad (20)$$

with the abbreviation $Z_o = \sqrt{M/C_d}$. As shown in section 2.5, tg φ is usually rather small and is independent of frequency since, as mentioned η_w decreases with increasing frequency. For the phase constant k we find

$$k \stackrel{\wedge}{=} \omega\sqrt{MC_d} \qquad (21)$$

Therefore, the phase velocity in a reflection free tube is

$$c = \omega/k \stackrel{\wedge}{=} 1/\sqrt{MC_d} \qquad (22)$$

and independent of the frequency. The damping coefficient is

$$\beta \stackrel{\wedge}{=} Z_o G/2 + ktg\varphi/2 \qquad (23)$$

Since k increases proportional to the frequency , the damping factor ß in a viscoelastic tube increases with frequency.

In small arterial vessels we can assume leakage G, the viscous wall resistance R_w and the effective mass M to be negligible compared to the other components (Gross 1977). Under these conditions we find as approximate solutions − C is the compliance of the wall:

$$\beta + jk = \sqrt{j\omega RC} = (1 + j)\sqrt{\omega RC/2} \qquad (24)$$

Damping as well as the phase constant increase with the fre-

quency : $\beta = \sqrt{\omega RC/2}$ (25)

                                  $k = \sqrt{\omega RC/2}$ (26)

Therefore also the wave velocity is rather small and increa-

ses with increasing frequency (Caro et al.1978; see sect.3.6).

                                  $c = \sqrt{2\,\omega/RC}$ (27)

The characteristic impedance turns out as a complex function

                          $Z = (1 - j)\sqrt{r/2\,\omega\,C}$ (28)

The amplitude of the impedance decreases with increasing fre-

quency, in contrast to the characteristic impedance of large

vessels. The phase angle is constant -45°. - It should be

noted that only one solution of the square root which permits

a positive real part is physically sound. Actually the charac-

teristic impedance measured in small vessels agrees well with

this prediction. Fig. 26 is taken from Wetterer et al. 1977

and shows the characteristic impedance of the femoral artery

of a dog.

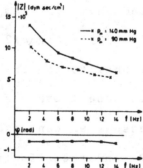

Fig. 26 Characteristic impedance of the femoral
artery of a dog (Wetterer et al. 1977)

Gross (1977) has compared measurements of the pressure amplitude in the microcirculation with results of a simulation in a model, which consists of 5 levels of dichotomic branching microvessels. Fig. 27 shows the calculated relative pressure amplitude in the vessels of this network (Gross 1977). The pressure decay is mainly due to the marked damping by viscous friction of fluid resistance.

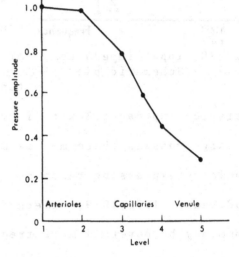

Fig. 27 Decay of pulsatile pressure amplitude in a network model of the microcirculation (Gross 1977)

3.3 The high frequency input impedance of arteries

In this connection "high frequency" means heart rate and its harmonics. Besides that the input impedance always comprises the zero frequency value, i.e. the mean absolute value of the peripheral resistance at the end of the examined arterial bed (R_{abs}). - As far as the pulsatile component of the pressure-flow relation is concerned, the input impedance also contains information about R_{diff} as explained in fig. 28 (Kenner 1978). The low frequency part of the impedance extends between the points R_{diff} and the heart rate.

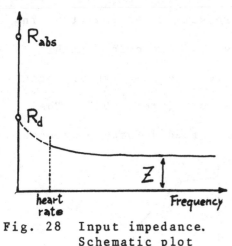

Fig. 28 Input impedance.
         Schematic plot

The contour of this part contains information about reactions of the arterial bed as will be discussed in part 4. In fig. 28 no low frequency reaction is assumed - the peripheral resistance is constant. The high frequency part of the input impedance of arteries tends to the value of the characteristic impedance Z with increasing frequency. Particularly in large vessels there may be maxima and minima of the input impedance expressing resonance phenomena (Wetterer and Kenner 1968,Kenner 1972,1979, Westerhof et al.1979). At least one minimum may be observed more frequently somewhat above the heart rate. For the simulation of this minimum a manometer model (see below) may be used. The input impedance of a homgeneous large tube with a resistance at its end shows several resonance maxima. The rather flat high frequency part of the arterial impedance in vivo is due to tapering and branching of the arteries (Kenner 1972). A recent summary about aspects concerning the arterial input impedance was presented by O´Rourke (1982).

Windkessel models have been used to simulate input impedance of the whole arterial system or of arterial branches. Two of the more important models are the "improved windkessel" model by Broemser and Ranke (1930) -recently sometimes called "westkessel" (Noordergraaf 1978) - and the manometer model

by Broemser (1932). The input impedance of the improved wind-

kessel model tends towards the series resistance R_1 (in the

nomenclature of fig. 29 which is taken from Wetterer and

Kenner 1968). This makes the model useful to simulate arteri-

al branches since R_1 simulates the effect of the characteris-

tic impedance Z (as indicated in fig. 28). The second model,

the manometer model includes the effect of the effective

mass and accounts for one minimum of the input impedance

which in some in vivo recordings can be observed (Westerhof

et al. 1979, O'Rourke 1982).

Fig. 29   Left: "improved windkessel" by Broemser and
          Ranke (1930). Right: manometer model by
          Broemser (1932)

There are two possibilities to plot the frequency depen-

dence of the input impedance. The usual way is the seperate

plotting of modulus and phase. Ronniger 1954 proposed locus

plots for the combined display of moduli and phases - which

has some advantages if one is used to the representation.

Fig. 30 shows a comparison between both plots . The example

of an actual measurement in a hydrodynamic elastic tube mo-

del is shown. The right upper part gives a simple explana-

tion for the construction of the locus plot from the modulus

(amplitude) and phase data. The input impedance of a non

branching tube with rather small damping shows a series of

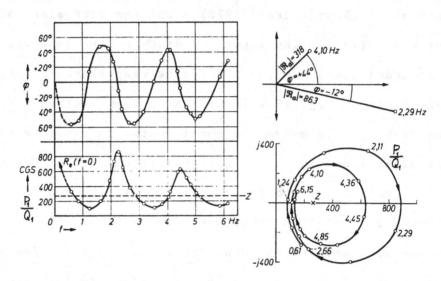

Fig. 30 Input impedance of a homogeneous tube. Left:
phase and amplitude, right: locus plot

maxima and minima due to resonance. For a comparison, in fig.
31 the locus plots of the input impedance of the two windkessel models from fig. 29 are shown. Highly characteristic
is the negative half circle which is passed through in clock-

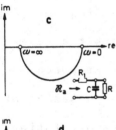

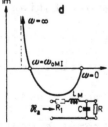

Fig. 31 Input impedance
of windkessel and
manometer

wise direction with increasing
frequency. Fig. 32 shows a recording of the input impedance of
the aorta of a dog during different experimental conditions in
order to demonstrate the rather
flat high frequency part. The
values of the zero frequency
(R_{abs} ,in the fig. 32 indicated
as Z_o) are written in the insert
(Kenner et al.1968).

For a comparison, finally a typical locus plot of the high

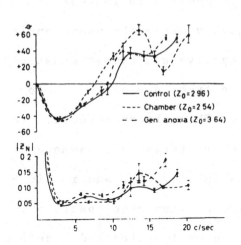

Fig. 32 Input impedance of
        ascending aorta of
        a dog (Kenner et al.
        1968)

frequency part of the input impedance of a dog is shown in
fig. 33. In this figure the
points of the absolute and differential resistance value are
indicated. The harmonics of the
input impedance are arranged in
a half circle which resembles
the half circles of the models
shown in fig. 31. - A corresponding locus plot including an in
vivo recorded low frequency

part is shown in fig. 47 in section 4.2 . One consequence of

the typical pattern of the high frequency part of the input

impedance is the fact, that the phase of flow always precedes

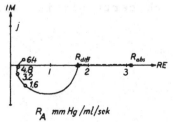

Fig. 33 Input impedance of
        ascending aorta of
        a dog (Kenner 1975)

the phase of the pressure. In
the early days of hemodynamic
modeling this fact was sometimes
called "Franks rule" in honour
of O.Frank (1899) who was the
first to study pressure and

flow in windkessel models.

    The input impedance of a small artery is shown in fig.

34 (Busse et al. 1979). Pressure and flow was recorded in a

small muscular branch. The modulus of the impedance decrea-

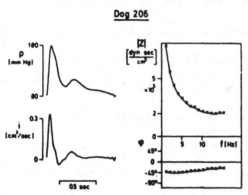

Fig. 34 Pressure and flow
(left) and input im-
pedance of a small
artery (right).From
Busse et al. (1979-
Abstract Euromech
118)

ses markedly with increasing
frequency while the phase angle
is negative and almost constant
in the frequency region up to
30 Hz. The pattern resembles to
the characteristic impedance
shown in fig. 26 , and is similar to a "windkessel pattern".
For the explanation and modeling
Busse et al. have assumed tapering of small vessels. However,
the influence of the characteristic impedance as discussed
above in section 3.2 should also be considered. - In any case
it seems important to note the "windkessel- like" properties
of the small peripheral arteries which certainly is important
for modeling the arterial system.

3.4 Pressure and flow transmission

As the arterial pressure and flow pulses are transmitted
along the tube system, both pressure and flow waves show
characteristic changes. An example from a linear transmission
line model (Kenner 1979) is demonstrated in fig. 35. This
figure shows from above to below: the central arterial pressure pulses,the arterial pulses along the system, the flow
pulses normalized with the local characteristic impedance.

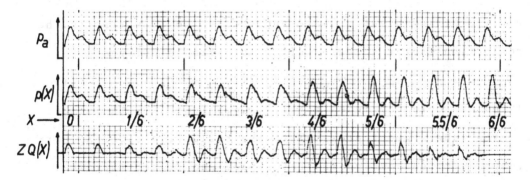

Fig.35 Pressure and flow along transmission line
model. Description in the text

In this model (Kenner 1965,1979) the main arterial line is

simulated by an inhomogeneous linear transmission line. The

tapering of the line is simulated by 3 reflections at the

locations 11,23 and 61 within a total length of 100 units.

The chosen reflection factors are 0.2,0.5 and 0.3; the re-

flection factor at the end of the tube is assumed as 1.0

corresponding total reflection. The outflow from the system

is assumed by leakage along the tube (side branches).The lea-

kage generates a damping of 0.006 per unit. The total increa-

se of the characteristic impedance from central to periphery

is 8-fold . In agreement with in vivo recordings we find the

following characteristics in fig. 35. The pressure amplitudes

increase from central to periphery. The contours of the pres-

sure pulses change in a very typical pattern, in that the

secondary (dicrotic) wave gets more marked. The flow ampli-

tude decreases from central to periphery (this is concealed

in fig.35 since the flow amplitudes are normalized by multi-

plication with the characteristic impedance). Towards the peri-

phery a marked negative wave appears in the flow tracing. The
reflection factor at each reflection site is defined as

$$k_n = (Z_{n+1} - Z_n)/(Z_{n+1} + Z_n) \qquad (1)$$

In our example the index n = 1,2,3,4,5 indicates the number
of the tube segment proximal to the reflection site. Z_5 corresponds to the peripheral resistance and is set infinite.

In the following we restrict ourselves to the pressure
transmission function (PTF) which is calculated by estimation
of the complex quotient (or, in other words, the modulus and
the phase angle of the quotient of the harmonic components)
of two pressure pulses

$$PTF = P_1/P_2 \qquad (2)$$

If pressures are recorded in a reflection free transmission
line there will be no change of the amplitudes of any frequency. The phase shift entirely depends on the local wave velocity. - If two pressures are recorded in a tube with reflections the pattern will be more complicated because the phase
velocity as well as the amplitudes are influenced by peripheral wave reflection. - In section 3.2 the pressure transmission function in a homogeneous tube with a resistance at its
end was theoretically derived. In order to demonstrate the
most simple case we assume now a friction-less flow so that
the damping coefficient ß = 0. Under this condition the
pressure transmission function can be written as

$$P_1/P_2 = \cos kL + j(Z/R)\sin kL \qquad (3)$$

where L is the length of the tube and the transmission func-

tion is estimated between entrance and peripheral resistance.

The equation shows, that the basic shape of the function is

an ellipse. A recording in the same homogeneous elastic tube

model and under the same condition as was shown in fig. 30 is

displayed in fig. 36. Here again the two possible ways of

plotting are shown.

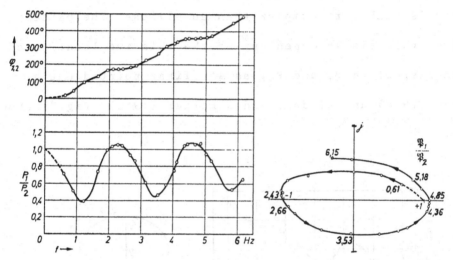

Fig. 36 Phase and modulus (left) and locus plot
(right) of the pressure transmission function P_1/P_2 in a homogeneous tube (Wetterer
and Kenner 1968)

The locus plot shows the characteristic ellipse, somewhat

distorted by the viscoelastic damping. The relation between

the two axes of the ellipse corresponds to the quotient

$$F = Z/R \qquad (4)$$

Therefore the pressure transmission function between the en-

trance and the end of a vascular segment is a very sensible

indicator for changes of the peripheral resistance R. Vaso-

dilatation (decrease of R) increases the ellipse to a circle
and vasoconstriction leads to a flattening of the ellipse.

It should be mentioned that tapering makes the ellipses
excentric as can be seen in all in vivo recordings. Furthermore, if the transmission function is measured between entrance P_1 of an arterial segment and some point P_x which is located proximally to P_2 then the pattern of the transmission
function is much more complex than an ellipse. The pattern is
highly characteristic dependent on the relative location of
P_x with respect to P_1 and P_2, as was first pointed out by
Ronniger (1954) and Lindner and Ronniger (1955). Fig.37 shows

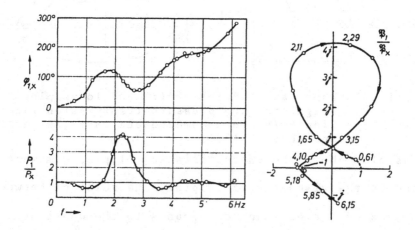

Fig. 37 Pressure transmission function P_1/P_x where
P_x was measured in the middle between P_1 and
the end of the tube (P_2). (Wetterer and
Kenner 1968)

a characteristic example from the same model as in figs. 30
and 35. In the example of fig. 37 the pressure transmission

function was measured between the entrance of the tube and
the middle of its length. The following equation explains the
way to analyze the distortion of the locus plot (Lindner and
Ronniger 1955). If one writes

$$P_1/P_x = (P_1/P_2)/(P_x/P_2) \qquad (5)$$

it is clear that both functions P_1/P_2 and P_x/P_2 are ellipses
since both describe the transmission from an entrance to the
end of a segment. The division of the two functions leads to
the rather typical pattern shown in fig.37 and in the following figures. The location of the loop of the locus plot is
characteristic for the location of P_x with respect to P_1 and
P_2.

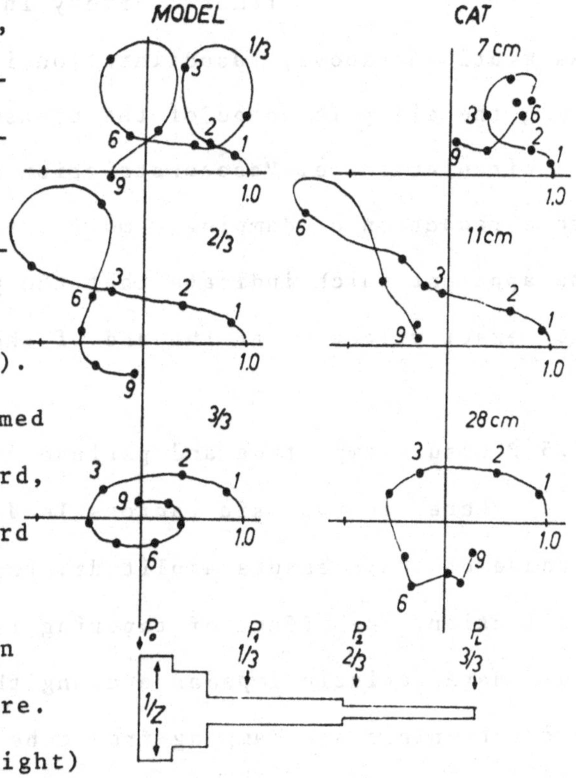

Fig. 38 (Kenner 1975,
1979) shows left the estimated pressure transmission function in the trans-
mission line model described in the beginning of
this section (Kenner, 1975).
The estimation was performed
from the entrance to 1/3 rd,
from the entrance to 2/3 rd
and from the entrance to
the end of the model shown
at the bottom of the figure.
The result is compared (right)

with measurements performed in a cat. In other animals and
in man simílar results have been obtained (Wetterer and
Kenner 1968). In the cat in which the transmission functions
shown in fig. 38 were measured, we have examined the effect
of vasoconstriction (NOR = noradrenalin) and vasodilatation
(ACCH = acetylcholin). The result id shown in fig. 39.

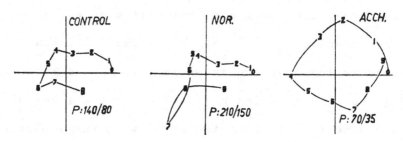

Fig. 39   Pressure transmission ascending aorta to
           femoral artery in a cat (Kenner 1979)

As mentioned above, vasodilatation is characterized by changing the elliptic shape of the transmission function to a more circular shape. Vasoconstriction narrows the ellipse. Due
to a reduction of damping, small "reflection-loops" may become apparent which indicate that the peripheral recording is
not exactly located at the end of the segment.

3.5 Pressure amplitude and peripheral pressure-flow relation

      There are two main factors leading to a peripheral in-
crease of the pressure amplitude: tapering and peripheral
reflection. The effect of tapering is due to the increase of
the characteristic impedance along the aorta (see fig.12).In
a frictionless and damping-free tube we would expect an asymp-

totic value of the peripheral amplification of the pres-

sure amplitude as follows (Kenner, 1972):

$$\Delta p_2 / \Delta p_1 = \sqrt{z_2/z_1} \qquad (1)$$

The second component of the pressure amplification is the pe-

ripheral reflection. At the site of the reflection we can ex-

pect $\qquad\qquad\qquad \Delta p_2 = \Delta p_1 (1 + k) \qquad\qquad (2)$

in a homogeneous tube; if Δp_1 is the amplitude of the incom-

ing pressure wave, and k is the local reflection factor, then

Δp_2 is the actual amplitude recorded.

Kenner and van Zwieten (1982) have discussed the impli-

cations of changes of the distribution of the pressure ampli-

tude along the aorta in relation to the peripheral reflection.

Fig. 40 shows the distribution of pressure amplitudes along

the aorta of 8 cats (A ascending aorta,M mesenteric artery,

B iliac bifurcation, F femoral artery). The total amplificat-

ion is more than 2-fold which is in agreement with combined

effects of tapering and reflection. We observed that vaso-

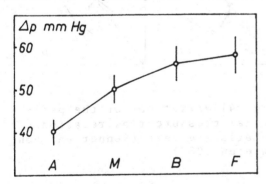

dilatation, produced by different drugs may lead to different changes with respect to the pressure amplification. We assume that a decrease of the

Fig. 40 Distribution of pressure amplitudes along the aorta.
       Details see text (Kenner and van Zwieten 1982)

peripheral resistance decreases reflection , and thus also,

amplification. This pattern can be actually found with acetyl-
choline. Other vasodilating drugs like e.g. hydralazin inter-
restingly lead to an increase of the pressure amplification.
This result shows that we are still far away from fully understanding the physics of the peripheral pressure-flow relation
in small arteries and arterioles.

The peripheral reflection factor k is defined as

$$k = (R_d - Z)/(R_d + Z) \tag{3}$$

where R_d is the differential value of the resistance. From
our observation of the possibility of discordant behaviour of
mean pressure and pressure amplitude (amplification) we may
consider that absolute and differential value of the peripheral resistance must not always change in the same direction.
In a highly schematized drawing we have tried to express this

in fig. 41. In both
examples shown in the
figure the absolute va-
lue of the resistance
is assumed to decrease
(shift to the left). In
the left diagram the
slope $(dq/dp = 1/R_d)$
of the pressure-flow re-

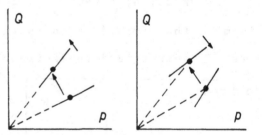

Fig. 41 Variations of the peri-
pheral pressure-flow relation
Details see text (Kenner and van
Zwieten 1982)

lation changes into the same direction. In the right diagram
the slope changes into the opposite direction. It will be

necessary, of course , to confirm this interpretation by ano-

ther method.

A further interesting aspect concerning the peripheral

pressure-flow relation is shown in fig. 42 which shows a de-

tail from fig.35: pressure and flow in the periphery of the

tube model (Kenner 1979). The

outflow from this model is ass-

umed to take place through small

side branches (leakage). The re-

flection factor at the very end

is one. So outflow at the end

is zero. A few units of the model

proximally to the end a flow

pattern can be observed which

closely resembles the time de-

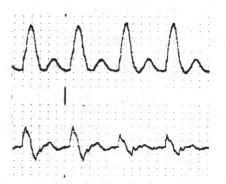

Fig. 42 Pressure (top)
and flow (bottom) in the
periphery of the tube mo-
del (Kenner 1979)

rivative of the pressure pulse. The record is also very simi-

lar to pressure and flow in a small artery (fig. 34). This

phenomenon leads again from a slightly different viewpoint to

the fact that the periphery behaves like a windkessel. In

other words, the pressure and flow pattern in figs.34 and 42

can, formally,be interpreted by the windkessel equation:

$$q \;=\; p/R \;+\; C\,dp/dt \tag{4}$$

which implies in the case of $k = 1$ and $R = \infty$:

$$q \;=\; C\,dp/dt \tag{5}$$

The impedance, therefore, is

$$P/Q = 1/j\omega C \tag{6}$$

and has a modulus which decreases with frequency and a negative phase angle for all frequencies of -90°, very similar to the pattern shown in fig. 34.

3.6 Wave velocity

The speed with which a signal (pulse) is transmitted through an artery depends on the following factors:

1) Elastic modulus of the vessel wall

2) Radius of the vessel

3) Wall thickness

4) Damping and viscosity

By influencing one or more of these factors the wave speed shows correlations with the following magnitudes or functional changes: pressure inside the vessel (better: transmural pressure difference). Contractile status of the vascular smooth muscles. Age and pathologic changes of vessel wall geometry and composition.

The wave velocity can be described by "Webers equation"

$$c = \sqrt{(dp/dr)(r/2\rho)} \tag{1}$$

In the case of small extensions the pressure radius relationship is described by Youngs modulus E. Insertion into the above equation yields "Moens-Korteweg" equation:

$$c = \sqrt{Eh/(2r\rho)} \tag{2}$$

In elastic tubes with large extensibility we have to use

an incremental modulus E_t as discussed in section 2.6. In
the following equation a force-related incremental modulus
is used (Wetterer and Kenner 1968, Kenner 1967):

$$c = \sqrt{(1/2\rho)(\frac{E_t h}{r(1 - \mu^2)} - p)} \qquad (3)$$

In these two equations the effect of friction and of viscous
energy losses is neglected.

In small arteris, arterioles and capillaries the wave
velocity is markedly influenced by fluid viscosity, as discussed in section 3.2. There the wave velocity was found to
be $\qquad\qquad\qquad c = \sqrt{\omega RC/2} \qquad\qquad\qquad (4)$

The wave velocity in large arteries has the same order
of magnitude in animals of different size. Towards the periphery as the content of muscle fibers in smaller arteries increases, the wave velocity increases as long as the effect
of the fluid viscosity is relatively small. In the dog the
following values of the wave velocity can be observed, according to Caro et al. 1978): central aorta 400 - 600 cm/s ,
abdominal aorta 600 - 750 cm/s, femoral artery 800 - 1000
cm/s. All these velocities may be higher if the arterial pressure increases (see below). In small vessels the influence
of the viscous resistance is marked and decreases the wave-
velocity. In capillaries values as low as 10 cm/s are reported (Caro et al. 1978).

Vasoconstriction increases the wave velocity by decreasing the radius. However, the question whether an increase

of smooth muscle tone ceteris paribus would increase or decrease the modulus of elasticity of the vessel wall is still
lacking a general answer. Probably in some cases the modulus
does not measurably change, in some other vessels a decrease
of the modulus is actually possible.

The elastic modulus of the vessel walls in any case increases if the arterial pressure rises. One example by Wetterer
and Pieper (1953) of the relation between arterial pressure
and wave velocity is shown in fig. 43. There is a general
trend of the wave velocity in the aorta to increase with increasing pressure. However, during Sympatol effect (Sy) the
wave velocity is shifted towards lower values indicating a
decrease of elastic modulus as a response to the sympathomi-
metic drug.

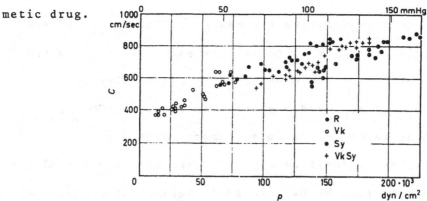

Fig. 43 Relation between arterial pressure and wave velocity
        in the aorta of a dog. R control, Vk vagus cooling
        Sy Sympatol injection, VkSy same during vagus cooling
        (Wetterer and Pieper 1953)

If in an artery the transmission of a pulse is measured several additional factors influence the observed result. Usually the transmission is measured at the pulse front at the

mean arterial pressure value or at 1/5 th of the pulse height.

Using this method the influence of reflections seems minimal

although it cannot be excluded completely.

A marked influence of reflected waves can be observed

if the phase velocities of harmonic frequency components are

measured from the pressure- or flow- transmission function.

The equation for the pressure transmission function (eq. 4

in section 3.4) shows that the phase transmission time for

the first harmonic frequency usually is much smaller than the

phase transmission time for higher harmonics. A typical exam-

ple from a cat is shown in fig. 44.(Kenner and van Zwieten

1982). The figure shows

the locus plot of the

pressure transmission

function before (full

line) and after the in-

jection of the vaso-

dilating drug hydrala-

zin (dashed line),

Two observations

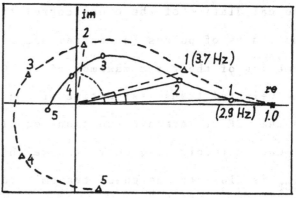

Fig. 44   Pressure transmission function
before and after vasodilatation
(Kenner and van Zwieten 1982)

are pertinent: The phase angle of the first harmonic is much

smaller than 1/2 of the phase angle of the second harmonic.

Thus, the phase velocity of the second harmonic is much smal-

ler than the phase velocity of the first harmonic. The phase

velocity is defined by the following equation:

$$c_{ph} = (\omega/\phi) \; x = (n \, \omega_1 / \phi_n) \; x \qquad (5)$$

ω is the frequency and ϕ is the corresponding phase angle. ω_1 is the frequency of the first harmonic of a pulse wave and n is the order number of the harmonic frequency. ϕ_n is the phase angle of the n-th harmonic frequency. In general the phase velocities of lower harmonics are higher than those of higher harmonics, due to the influence of reflections. During vasoc dilatation and decrease of the blood pressure both the transmission times and the phase angles increase - and the corresponding velocities decrease.

A worthwhile method for measuring transmission times is the calculation of the cross correlation function (CCF) of two trains of pulses (a central and a peripheral one). The time lag of the first maximum of the CCF corresponds to the weighted transmission time of the pulses (Kenner 1979). As the CCF of harmonic functions equals the sum of the CCFs of each harmonic frequency, the weighted transmission time τ_{CCF} is close to the phase transmission time of the first harmonic which is always the largest component of the arterial pulse. If the CCF is calculated after differentiation of the pulses with respect to time then the transmission time τ_{CCFD} corresponds to a weighted transmission time of higher harmonics. Fig. 45 shows the dependence of τ_{Front}, τ_{CCF} and τ_{CCFD} on arterial blood pressure in a rabbit and a cat. The blood pressure variation was brought about by infusion of noradrenalin. In agreement with the pattern demonstrated in

fig. 44 the transmission times for lower harmonics depart

from those of higher harmonics the more the higher the blood

pressure and the more pronounced the peripheral vasoconstric-

tion.

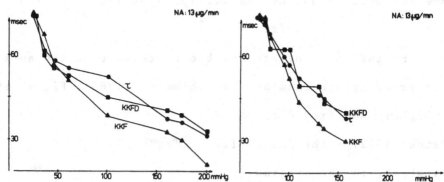

Fig. 45   Transmission times measured along the aorta of a
cat (left) and a rabbit (right) using three different methods:
τ propagation measured at the front of the pulse wave.KKF:
using cross correlation. KKFD: same after differentiation.

It can be proposed that the continuous recording of trans-

mission times or wave velocities can be used for the assess-

ment of arterial blood pressure. - In order to summarize the

definitions of different velocities the following equations

are presented: The phase velocity without reflections

$$c = \omega/k \qquad (6)$$

The group velocity in absence of reflections

$$c_g = d\omega/dk \qquad (7)$$

Due to the rather unique skewed distribution of harmonics in

pulse waves, the group velocity of arterial pulses does not

play a large role in practical considerations. The phase ve-

locity $\qquad c_{ph} = (\omega/\phi)\Delta x \qquad (8)$

as mentioned above. Finally, the apparent phase velocity at
a certain location is defined as

$$c_{app} = \omega dx/d\phi \tag{9}$$

The signal velocity is influenced by the flow velocity in a
tube (v) and, as Anliker et al.(1968) proved, corresponds to

$$dx/dt = c \pm v \tag{10}$$

The influence of age and blood pressure on the aortic
wave velocity was examined by Schimmler (1965). Fig.46 shows

a slightly modified diagram
(Kenner 1972) - the dashed lines which, in the original
are distorted due to selecti-
on by death of the patients,
were straightened. The aortic
wave velocity in humans is shown
as a function of age and blood
pressure. The wave velocity was
measured from A.carotis to the
femoral artery in 2500 persons.

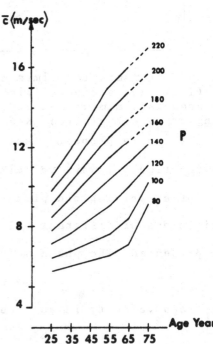

Fig. 46 Aortic wave velocity
as a function of age and mean
arterial pressure. Data from
2500 persons. After Schimmler
1965.

4. SMALL ARTERIES AND THE CONTROL OF FLOW AND PRESSURE

4.1 Introduction

We have in part 3 introduced the concept of frequency dynamics. The low frequency part - below the heart rate - represents effects of slow reactions of the living vessel wall to changes of pressure and flow. The more pronounced and interesting reactions in this frequency region are active and related to metabolism. The reactions are initiated by autoregulation (local effects to changes in flow and pressure) or by neural control which, in most cases , is centrally coordinated. The low frequency reactions also comprise passive effects as slow stress relaxation and creep.

In the small vessels all the influences which possibly can change pressure and flow come together as indicated in fig. 47 after Folkow and Neil 1971. Here also the interactions take place between extrinsic influences like drugs and neural and autoregulatory feedback mechanisms.

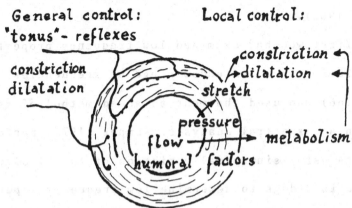

Fig. 47   Summary of the regulatory influences in the region of the small arteries.

4.2 The low frequency input impedance

All the discussions in the preceding sections were some-
how related to the question how the properties of small resistance vessels influence pressure and flow in the circulation.
The pressure-flow relation in the resistance vessels was discussed as a nonlinear function (2.8) and as related to the
characteristic impedance of small arteries (3.2).Furthermore,
we have attempted to distinguish absolute and differential value of the resistance (3.3) and the upstream influence of the
resistance due to reflections (3.4 to 3.6). - The method by
Wetterer and Pieper (1953) to determine the pressure-flow relation of the peripheral resistance used pressure and volume
variations artificially induced by a pump with a frequency
of around 0.2 Hz. - The question now of course arises whether
this particular frequency shows a particular and different
behaviour than other frequencies in the region below the
heart rate (Kenner 1971,Attinger 1973,Basar 1981,1982, Sipkema et al. 1982).

The first who had examined low frequency properties of
the circulation in an extended range of frequencies was M.G.
Taylor (1966) who used the most advanced method of random
excitation and spectral analysis. Kenner (1971) performed
experiments using sinusoidal variation of the blood volume
of anesthetized dogs to determine low frequency input impedance (the complex pressure-flow relation) of the aorta.

Fig.48 shows a slightly schematized comparison of the aortic
input impedance of a dog with the low frequency part display-
ing regulatory phenomena (upper part) and without, i.e. after
paralysis of vascular reactivity.

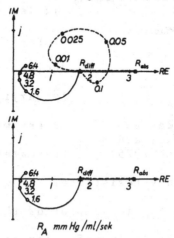

Fig. 48   Input impedance of the aorta of a dog.
           Upper part including regulatory pheno-
           mena. Kènner 1975)

With respect to the small vessel reactivity, a futher important step was the application of sinusoidal perfusion in order
to examine the low frequency impedance of different arterial
beds which was performed by Basar and coworkers. Basar rec̄
cently summarized his results in a book(Basar 1981,1982).
Rubenstein et al. (1973) used pseudorandom binary sequences
(PRBS) to test and analyze local hemodynamic control.
Following these studies Kenner and Ono (1971) and Kenner
and Bergmann (1975) have examined the low frequency input
impedance of different arterial beds. Fig. 49 shows the locus plot of the input impedance of the femoral artery of a

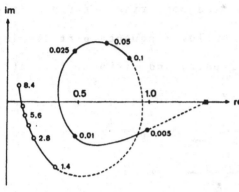

Fig. 49 Input impedance of
the femoral artery of a dog
(Kenner and Bergmann 1975)

dog. The circles show the
high frequency impedance
part, measured from pressure
and flow pulses. The filled
circles show the low frequency part measured with introducing artificial sinusoidally varying perfusion of the
artery with blood. The square
indicates the absolute value
of the zero frequency resistance.

Fig. 50 shows the locus plot of the low frequency impedance of the renal artery of a dog at three different per-

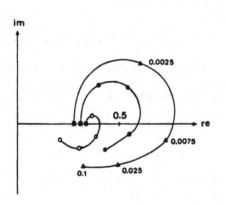

Fig. 50 Input impedance of
the renal artery of a dog.
Low frequency part at three
different mean pressures
(Kenner and Bergmann 1975)

fusion pressures. Circles: 80
mm Hg, filled circles: 150
mm Hg , triangles: 200 mm Hg.
This figure indicates that
the reactions are pressure-
dependent and, in a wider
range of pressure, nonlinear.

Trying to explain in
short these impedance pattern
we have to assume that the
interaction between pressure

and flow represents regulatory phenomena: in local peri-

pheral beds autoregulation. In the general circulation, represented by the aortic input impedance, in addition reflex
reactions, particularly baroreceptor reflexes, play an important role.

Autoregulation in the small resistance vessels has the
following physiological correlates. 1)Metabolic effects: if
the local flow decreases the supply with oxygen and the transport of metabolic products is reduced. These changes will,
directly or indirectly dilate vessels in order to increase
flow and vice versa. 2) Myogenic autoregulation: if the vessel wall is quickly stretched, a contractile response may be
observed (Bayliss effect). Both these effects will be summarized below as "contractile response C_q" to increase of flow.
3) Autoregulation of arterial pressure. Similar to the reaction of the baroreceptor sytstem there seems to be a local
reactive "vasodilatation response D_p" to pressure increase
(Kenner and Ono 1971).

We describe the resistance R of the small vessels as
a function of pressure and flow.

$$p = q R \qquad\qquad (1)$$

and
$$R = R(p,q) \qquad\qquad (2)$$

In order to derive a linear approximation we use the following procedure which recently has also been used by Hatakeyama (1982). Differentiation of the two equations yields

$$dp = q dR + R dq \qquad\qquad (3)$$

and $\qquad\qquad dR = (\ \partial R/\partial p)dp + (\ \partial R/\partial q)dq$ $\qquad\qquad$ (4)

and after insertion

$$\frac{dp}{dq} = R \ \frac{1 + (q/R)(\ \partial R/\partial q)}{1 - (p/R)(\ \partial R/\partial p)}$$ (5)

The two factors in the right side can be interpreted as gains of the resistance control system; The following two functions are defined: the gain of flow control

$$C_q = (q/R)(\ \partial R/\partial q)$$ (6)

As explained above C_q is supposed to represent contractile responses to flow increase. The gain of the pressure control

$$D_p = - (p/R)(\ \partial R/\partial p)$$ (7)

The negative sign takes into account that an increase of the pressure dilates the vessel and thus decreases the resistance. The two effects C_q and D_p are reciprocal as far as the influence on the resistance is concerned (Kenner and Ono 1971). A more general formulation including dilating effects of flow and constricting effects of pressure was given by Kenner (1971). For practical applications the way shown here seems sufficient for complete description of phenomena.

We now assume that the gains can be expressed as frequency dependent functions and chose as simplest possibility first order transfer functions (in Laplace formulation):

$$C_q = C/(1 + s\tau_q)$$ (8)

and $\qquad\qquad - D_p = D/(1 + s\tau_p)$ $\qquad\qquad$ (9)

Here we write the Laplace frequency operator s instead of

the Fourier operator j   in order to express the necessity
to calculate solutions in the time domain. Using these functions, the low frequency impedance can be written as

$$\frac{dp}{dq} = R \frac{1 + C/(1 + s\tau_q)}{1 + D/(1 + s\tau_p)} \qquad (10)$$

C and D are dimensionless constant values of the gains. τ_q
and τ_p are time constants. R is the (mean) zero frequency
value of the impedance. (R actually contains information about the values R_{abs} and R_d and about the high frequency part
of the impedance. Fig. 51 shows a comparison of calculated
locus plots (left) and locus plots measured in different
arterial beds of a dog (Kenner and Ono 1971).

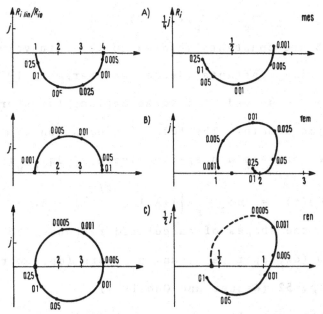

Fig. 51 Low frequency impedance. Left: calculated. Right:
        measured. mes   mesenteric artery, fem femoral arte-
        ry, ren renal artery. Description see text.
        (Kenner and Ono 1971)

A)Left: Pure autoregulation of flow, $D_p = 0$, $C_q = 3/(1 + 5s)$.

Right: low frequency impedance of the mesenteric artery.

B) Left: Autoregulation of pressure, $D_p = 3/(1 + 100s)$, $C_q = 0$.

Right: Low frequency impedance of the femoral artery.

C) Left: Combined autoregulation of pressure $D_p = 3/(1+1000s)$

and flow $C_q = 3/(1 + 5s)$. Right: low frequency impedance of

the renal artery.

It can be shown by partial fraction expansion or by in-

sertion that the equation for the low frequency impedance

(eq.10) can also be written in the following form:

$$\frac{dp}{dq} = R (1 + A/(1 + s\tau_q) - B/(1 + s\tau_p^*)) \qquad (11)$$

where

$$\tau_p^* = \tau_p/(1 + D) \qquad (12)$$

and A and B are constants related to the gain values C and D

and to the time constants (Kenner and Bergmann 1972). The

equation can be solved by inverse Laplace transformation

into the time domain. The following equation describes the

pressure response to a step flow increase dq_{step}:

$$(13) \qquad dp(t) = dq_{step} R\left[1+A(1-e^{-t/\tau_q}) - B(1-e^{-t/\tau_p^*})\right]$$

Examples of the shapes of calculated responses to step chan-

ges to flow (di,upper part) and pressure (dp,lower part) is

shown in fig. 52 (Kenner and Ono 1971).

From these results we can conclude that reactions of

vessels to changes in pressure and flow can be described by

a combination of pressure and flow control gains. The frequency region of these metabolic and myogenic control reac-

tions extends down from the heart rate towards about 0.001 Hz

or even lower (there may be a smooth transition towards cir-

cadian rhythms or even longer periods). We did not discuss

here in detail the region of nervous mechanisms as barorecep-

tor reflexes which are most pronounced in the frequency re-

gion between 0.01 and 1 Hz.

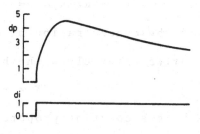

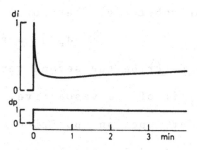

Fig. 52 Calculated responses
using eq. 13. Upper part pressure response (dp) to a flow
(di) step. Lower part: flow
resonse to a pressure step.
Kenner and Ono(1971)

4.3 Humoral autoregulation

Flow in arteries carries information and humoral signals

especially to the small vessels. For a substance with very

slow decay rate we can expect the concentration in the blood

to be rather independent of flow. For a substance whose de-

cay rate is fast compared with the flow rate the concentra-

tion on the site of action becomes dependent on flow. Kenner

and Ono (1972) have described experiments to demonstrate a

possible mode of autoregulation based on these considerations. If a vasoactive substance is secreted intò an artery at a constant rate, the concentration X at the site of action is inversely proportional to the flow. We assume that the resistance some distance downstream from the site of pooduction of the substance is influenced by its concentration X:

$$R(X) = p/q \tag{1}$$

Differentiation leads to

$$(\partial R/\partial X)dX = dp/q - p \, dq/q^2 \tag{2}$$

and, if the concentration of the substance in the small arteries varies inversely with the flow we can write

$$dX = - k \, dq \tag{3}$$

where k is a constant proportional to the rate of production of the substance. Therefore

$$dp/dq = R(1 - k \, (\partial R/\partial X)(q/R)) \tag{4}$$

If we describe the second term on the right side as the control gain of the vasoactive substance and chose a first order transfer function - positive C_x for vasoconstriction, negative D_x for vasodilatation , e.g.

$$C_x = C \, (1 + s\tau_x) \tag{5}$$

we find by inverse Laplace transformation the pressure response to a flow step

$$dp(t) = R \, dq_{step}(1 - C \, (1 - e^{-t/\tau_x})) \tag{6}$$

This equation indicates that the pressure in the arterial bed increases immediately after the flow step and then declines back to its former value (during constant flow per-

fusion) because of the reduction of the resistance by the
dilution of the vasoactive substance in the increased volume
flow.

A similar reaction may be found for the effect of a
pressure step on local flow if the vasoactive substance is
vasodilating. Then the solution in the time domain is

$$dq(t) = (dp_{step}/R)(1 - \frac{D}{1+D}(1 - e^{-t/\tau_x^*})) \quad (7)$$

The equation describes the following phenomenon: if, under
the presence of a continuous secretion of a vasodilating substance the perfusion pressure is increased, the flow follows
with a transient increase and then exponentially declines
back to the control value. If the vasodilating gain is close
to one , there is a perfect autoregulation of flow - i.e. the
flow is kept constant under these conditions.

Flow regulation of this kind may be possible in vascular areas where vasoactive substances are secreted, like
renin in the afferent arterioles in the kidney, or where
substances are generated by local effects, like certain
prostaglandines during the process of thrombosis.

4.4 Instability of flow and pressure in small arteries

Using the model described in the preceding sections
Kenner and Ono (1972) have examined the interaction of flow
and pressure in the course of drug induced reactions. On the
one hand we were able to quantify the interaction of the

of the effect of vasoactive drugs and the baroreceptor reflex. On the other hand we observed oscillations of pressure
and flow as shown in fig. 53, particularly during the infusion of the vasodilating substance acetylcholin in dogs.

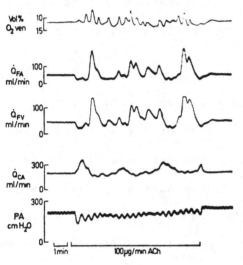

The tracings show from top to
bottom: femoral venous oxygen
saturation, femoral arterial
flow, femoral venous flow,
flow in the common carotid
artery, systemic arterial
pressure. The period of intravenous infusion of acetylcholin (ACh) is indicated at
the bottom. During infusion

Fig. 53 Autooscillations
during acetylcholin- infusion
in a dog (Kenner and Ono 1972)

of acetylcholin slow oscillations of the femoral flow
and, at the same time higher frequent waves (Mayer waves)
can be observed in the arterial pressure tracing. All these
observations indicate the generation of instability of flow
and pressure by the infusion of the drug.

An interpretation was attempted by the application of
the following equation of the low frequency impedance:

$$dp/dq = R(1 + C/(1 + s\tau_c) - D/(1 + s\tau_d))\qquad(1)$$

with C and τ_c gain and time constant of a vasoconstricting
effect (neural control or autoregulation). D and τ_d are gain
and time constant of the vasodilating influence of the in-

fused sub&tance. - If the magnitudes of both gains C and D

are equal and if the time constant of the dilating mechani-

sm is smaller than the time constant of the constricting me-

chanism, then instability may occur at the frequency

$$\omega_{in} = \sqrt{1/\tau_c \, \tau_d} \tag{2}$$

At this frequency small random pressure oscillations may

start "infinite", i.e. very large flow oscillations because

the low frequency input impedance (eq.1) tends towards zero

at this frequency. - We therefore assumed, that acetylcholin

may trigger a change of the time constant τ_d which then

starts the oscillations as shown in fig.53.

4.5 Autooscillations and vasomotion

As discussed by Basar (1981) it is certainly a mistake

to assume that all oscillations in the cirulation may be

explained by instabilities of local or general (neural) con-

trol. The smooth muscles of the vascular tunica media of

most small arteries has an immensely dense innervation which

generates an adrenergic vasoconstrictor tonus. Tonus is a

very fuzzy term which indicates contractile status depending

on the conditions: isometric stress if shortening is not

possible, or, isotonic shortening if constant stress is gi-

ven. In any case sympathetic activity under normal resting

conditions accounts for about 1/2 of the vascular smooth mus-

cle tone as can be judged from the decrease of vascular re-

sistance after paralysis of the sympathetic nerves - resistance falls to about 1/2 after such a procedure (by the way another fuzzy expression of the meaning of the term "tonus" of vascular smooth muscle).

If, during muscular exercise the resistance of small vessels in the muscles decreases to 1/10 or even less of its resting value, vasodilating metabolites and autoregulation play a major role besides the decrease of the sympathetic innervation. At the same time about 65 to 75% of the existing capillaries open up which have been closed or collapsed during rest (Folkow and Neil 1971).

The smooth muscles of probably all vessels have the capability to produce spontaneous excitations and contractions. In small vessels and arterioles and in precapillary sphincters the "vasomotion" may periodically occlude the lumen completely. The frequency of these contractions is reported between 10 and 60 per min (Noordergraaf 1978).

Since parallel vascular areas exhibit a synchronous vasomotion, autoregulatory phenomena may be described and explained as a statistical process. The frequency of the oscillations and the duration of closure varies under different local conditions in such a way as to provide autoregulatory control as described in section 4.2.

In larger vessels the autooscillations are related to contractile responses. Fig. 54 shows an example of a myogenic response of an isolated small vessel (rat tail artery) to a change of the internal pressure (Busse et al. 1982).

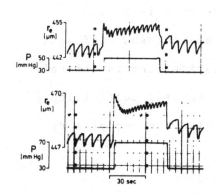

The small artery exhibits spontaneous activity. As the pressure is increased the frequency of the oscillations rises. At the same time the radius of the vessel undergoes a transient reaction, the shape of which corresponds closely to the double exponential function mentioned in section 3.7 and shown in

Fig. 54 Myogenic response of a small artery to a pressure step. (Busse et al. 1982)

fig. 52 - it is obvious that the reaction of the radius follow a similar contour as the reaction of flow to a pressure step.

5. THE DOWNSTREAM EFFECTS OF SMALL ARTERIES

5.1 Introduction

We do not really know why in all mammals the arterial blood pressure has the same order of magnitude (Kenner 1979). We know that the pressure decay from its arterial mean value of about 100 mm Hg (13.3 kPa) to the capillary pressure which is in the order of 15 to 30 mm Hg is due to the flow resistance of small arteries, arterioles and precapillary sphincters. The average value of the capillary pressure

is found to correlate with the value of the colloid osmotic
pressure of the plasma. In mammals the relation between arterial pressure and colloid osmotic pre sure is 5/1.

Since all the important exchange processes of fluids
and metabolites take place in capillaries, the functional
aim of the small vessels is directed towards controlling capillary flow and pressure, and thus also, transcapillary
fluid exchange.

5.2 Relation between resistance vessels and microcirculation

The small arteries, arterioles and precapillary sphincters act together as precapillary "arterial" resistance R_a.
We have already discussed the main influences which act upon
the state of contraction of the smooth muscles in the walls
of these vessels. The resistance vessels open into the network of capillaries. These exchange vessels are tubes formed
by a single layer of endothelial cells and a basement membrane, and have no capability of contraction. The blood from the
capillaries flows into the venoles and the small veins, vessels with some smooth muscles in their tunica media so that
a certain variability of this postcapillary "venous" resistance R_v can be anticipated. R_v is smaller than R_a and does
not contribute much to the total peripheral resistance. However, both R_a and R_v are important since the ratio of these
two resistances determines the magnitude of the hydrostatic

pressure p_c in the capillaries. The latter can be calculated
if the arterial pressure p_a and the venous pressure p_v are
given using the equation (see Folkow and Neil 1971)

$$p_c = \frac{p_a(R_v/R_a) + p_v}{1 + R_v/R_a} \qquad (1)$$

Besides diffusion and active transport by pinocytosis there
is a fluid flux through the capillary wall which is related
to the gradient of hydrostatic pressures in the capillary p_c
and in the surrounding tissue p_t, and the corresponding gradient of colloid osmotic pressures in the inverse direction.
π_t is the colloid osmotic pressure in the tissue spaces, π_{pl}
is the colloid osmotic pressure in the plasma. k is the filtration coefficient, S is the filtration surface area of the
capillary wall. σ is the socalled reflection coefficient of
the capillary wall with respect to a certain solute. Molecules which cannot pass the pores of the wall are reflected
with a factor of $\sigma = 1$. Under this condition the osmotic
pressure gradient can exert its maximum effect.

The flux Jf through the capillary walls into the tissue
space can be calculated according to the classical socalled
Starling hypothesis (Lee and Kenner 1982):

$$Jf = kS (p_c - p_t - \sigma(\pi_{pl} - \pi_t)) \qquad (2)$$

Whenever the sum of these terms on the right side is positive then filtration occurs and vice versa. It can be assumed
that under steady state condition mainly filtration occurs
in an amount which corresponds to about 0.1% of the plasma

inflow into the capillaries. Most of the filtered fluid is
recently assumed to enter the lymph vessels (Intaglietta 1977).

However, it should be mentioned that the numbers about
the absolute and relative values of the transcapillary flux
vary from tissue to tissue. The classical viewpoint according
to the Starling hypothesis assumes that filtration occurs
at the arterial end of the capillary and that reabsorption
is found at the venous end of the capillary and at the venole.
Thus, a "paracapillary flow" can be assumed which supposedly
amounts to about 1% of the plasma inflow to the capillary.
The difference between filtered and reabsorbed fluid enters
the lymph. - In non steady state, e.g. during changes of the
capillary pressure or during transition of osmotically hypertonic fluids through the
capillary net, there is no
doubt at all that reabsorption may occur. A schematic
drawing which explains the
phenomena in the microcirculation is shown in fig. 55

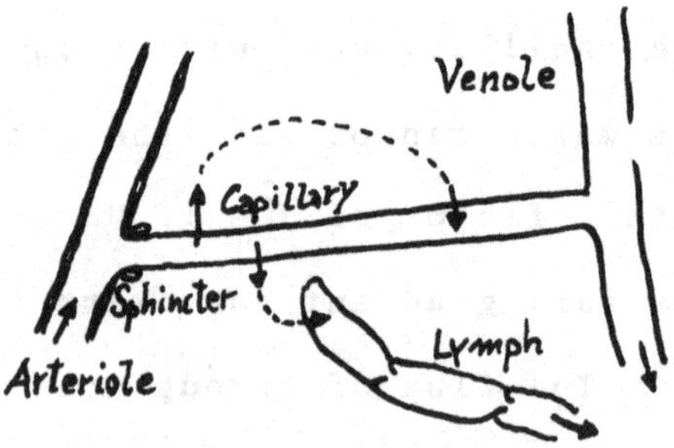

Fig. 55 Fluid exchange in
the microcirculation

5.3 Hematocrit in small arteries, arterioles and capillaries

One of the surprising observations about the properties
of the microcirculation is the extremely low hematocrit. Fig.
56 from Lipowsky et al. (1980) shows the hematocrit as a

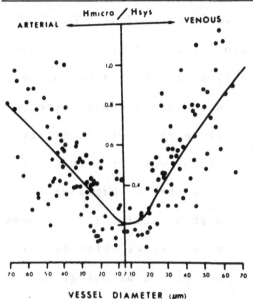

Fig. 56 Microvessel hematocrit as a function of the vessel diameter. Lipowsky et al. 1980

function of the vessel diameter: the smaller the vessel, the smaller the value of the hematocrit in comparison to the "feed" hematocrit in the large vessels (Klitzman and Duling 1979, Lipowsky et al. 1980). Other authors have made similar observations as summarized by Lee and Kenner 1982. - Thus, the microvessel hematocrit may be as low as 3% ,given a normal large vessel hematocrit of 45%.

The described phenomenon can most clearly be seen in resting muscle. During stimulation and activation of the muscle, the flow as well as the hematocrit increased in the capillaries to values close to the large vessel"feed"hematocrit. The observation that increase of flow through a vascular area simultaneously leads to an increase of the microvascular hematocrit seems very important for the function of the microcirculation.

Four, more or less interconnected, reasons may be cited for the interpretation of the effect (Fung 1981, Lee and Kenner 1982) : 1) Shunt flow of blood with high hematocrit through some thoroughfare channels other then capillaries. 2) Reduction of capillary hematocrit due to the entry condition at branching sites. As shown in fig. 57 from Fung (1981) the erythrocytes tend to move into the faster channel at branchings. In other words, the more flow, the more erythrocytes move into a small vessel.

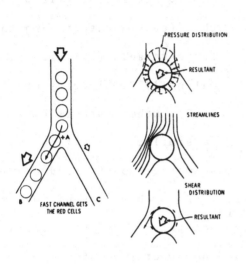

3) Decrease of hematocrit at the wall of a larger tube where the capillary siphons off the blood leads to an effect which is called plasma skimming. It can be observed that erythrocytes tend to move away from the vessel walls, so that a cell free plasma layer is generated.

Fig. 57 The branch with the faster stream gets the red blood cells. Fung (1981)

4) Connected with this observation is the Fahraeus-Lindqvist effect. The red cells in the core stream of the small vessels move faster than the plasma along the wall.

For all these reasons, the hematocrit in the small vessels is reduced. At the same time also the viscosity of the blood in the capillaries is low. However, as Lee and Kenner (1982) have reported, the explanation of the low hematocrit, including the search for possible artifacts, is still unsatisfactory.

5.4 Transport through the microcirculation

If an indicator solution is injected into some vessel
the solution is dispersed as it moves along the vessel, 1) by
convection, particularly by the behaviour of the velocity profile of the fluid; 2) by diffusion and 3) by splitting of the
stream through branchings of the vessels. The effects of the

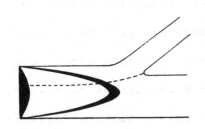

profile development and the influence of the side branch on
the distribution of an indicator
is schematically shown in fig.
58 from Lee (1980). He explained

Fig. 58 Development of an
indicator profile in a tube
at a branching point.
Lee (1980)

that different parts of the profile reach the branching points
at different times. Furthermore,

the passage times are different in different branches so that
at the venous end of a vascular bed the dispersion of the indicator can be found as mentioned above.

The effect of diffusion is marked in small tubes, where
the product of velocity and diameter is small compared to the
diffusion coefficient (Taylor 1953).

Lee (1980) has described the effects of convection and
diffusion in socalled "stream tubes". A stream tube is a func-
tionally defined transport unit which can be described by an
appearance time t_a and a transfer function h. Fig.59 from Lee
(1980) shows the transport function (=transfer function h of
the stream tubes) for the arterial, capillary and venous seg-

ment of a stream tube. The capillary part of the transport
function is markedly narrowed by the noticeable mixing effect
of the "Taylor diffusion" (Taylor 1953). Due to this diffusion the indicator profile front in a capillary is nearly flat
so that paradoxically the dispersion is minimized.

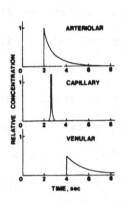

The overall effect on transport by the microcirculation can
be described through convolution
of the three functions shown in
fig. 59, and further by convolution of the result with the distribution function of the transit

Fig. 59 Transfer function
of the arteriolar, capillary and venular streamtube
segment of the microcirculation (Lee 1980)

times through all the branches
(i.e. all the parallel stream tubes) of the examined vascular seg-

ment. As an approximate description of the downstream transfer function corresponding to the transient on the venous
side after an arterial impulse injection of an indicator can
be given by the following simplified exponential function
(Lee and Kenner 1982):

$$h = 1(t - t_a)(\beta/t_a)e^{-\beta(t-t_a)/t_a} \qquad (1)$$

$1(t - t_a)$ is the unit step function starting at appearance
time. Using this function we also describe the arterial transient of an indicator after intravenous injection and passage
through the lung circulation. In this particular case which

will be discussed below with respect to the continuous recor-

ding of blood density, the overall mixing in the heart and

in the lung circulation is summarized by the described appro-

ximate equation (eq. 1).

5.5 A new method and its application for the study of the

    microcirculation

    Kenner et al.(1977) have introduced a method for the con-

tinuous recording of the blood density into physiological

and clinical research. As summarized in Kenner et al.(1982)

the technique is based on the mechanical oscillator principle

by Leopold et al. (1977).

    The physiological use of this technique is based on the

differences between the densities of blood and its components

and on the fact that the mechanical oscillator permits the

continuous recording of density with an accuracy of 10^{-5} g/ml.

The assumption is made and, under the most physiological con-

ditions can be proved, that the following linear mixing equa-

tion is valid for biological fluids including blood as sus-

pension of erythrocytes in plasma:

$$\bar{\rho} = \sum \rho_i v_i / \sum v_i \qquad (1)$$

The density of the mixed fluid $\bar{\rho}$ can be calculated from the

densities ρ_i and the volumes V_i of its i components. The fol-

lowing table shows the densities of some important components

of the blood and of solutions entering the blood or used for injection as test solution.

        erythrocytes   1085 - 1095 g/1

        plasma   1015 - 1020 g/1

        whole blood   1035 - 1055 g/1

        interstitial fluid   1000 - 1005 g/1

        isotonic NaCl (37°C)   1000 g/1

    20% hypertonic mannitol (37°C) 1061 g/1

Isotonic NaCl corresponds to an osmolar concentration of 310

mosmol/1. Hypertonic 20% mannitol corresponds to 1110

mosmol/1.

    The injection of isotonic solutions can be used as any

other indicator dilution method for the determination of car-

diac output and blood volume (Kenner et al. 1980).Furthermo-

re, the injection of hypertonic solutions can be used to de-

termine the filtration properties of the microcirculation.

During the passage of the hypertonic solution through the

capillaries fluid will be osmotically shifted from the extra-

vascular space towards the blood.

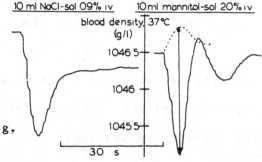

Fig. 60 Density transients
in arterial blood of a dog.
Details see text,
Kenner et al.(1982)

Fig. 60 shows the arterial density transient observed in the

carotid artery of an anesthetized 20 kg dog after intravenous injection of 10 ml of isotonic NaCl solution (left).
The transient is quite similar to dye dilution or to thermo-
dilution transients.

Once the dilution transient of an injected solution of
known density has been recorded, it is possible to calculate
the expected transient $\Delta\rho_{exp}(t)$ for any other solution with
known density. In fig. 60 the calculated expected density
transient after intravenous injection of 20% hypertonic mannitol solution is shown as a dashed line (right). - The actual observed transient $\Delta\rho_{obs}(t)$ is "distorted" by the osmotic influx of extravascular fluid (Kenner 1980,1982). The
observed transient thus is composed of two parts written on
the right side of the following equation:

$$\Delta\rho_{obs}(t) = \Delta\rho_{exp}(t) + \Delta\rho_{osmot}(t) \qquad (2)$$

A model of the process has been presented by Lee and
Kenner (1982). As discussed in section 5.4 the transient of
the injected solution can be described by a simple exponential function. After intravenous injection of a hypertonic
solution the bolus passes through the lung and appears in
the artery as modeled in the upper part of fig. 61 as
"hypertonic disturbance" which is the same as $\Delta\rho_{exp}(t)$ in
eq. 2 (Kenner et al. 1980).

The passage of the "hypertonic disturbance" leads by its
osmotic pressure to a fluid flux from the lung to the blood.
This fluid flux is shown in the lower part of fig. 61.

The flux corresponds to the time derivative of the weight
of the lung as measured in a gravimetric experiment. This
influx of extravascular fluid, furthermore is proportional
to the "osmotic density transient" $\Delta \varrho_{osmot}(t)$ and shows the
characteristic biphasic shape which was also observed in
the original recording of fig. 60. The biphasic shape of

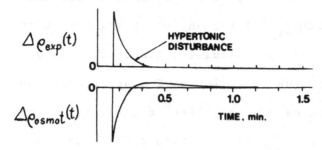

Fig. 61   Hypertonic disturbance and osmotic density
          transient according to a model by Lee and
          Kenner (1982)

the density transient can be explained by the fact that
the total mass of the solutes in the lung tissue does not
change. During the transition of the hypertonic bolus, therefore, the lung tissue through the extraction of fluid transiently becomes hypertonic. After passage of the bolus a reversed flux into the tissue transiently increases the blood
density.

     These experiments permit the determination of the trans-
port and the filtration properties of small vessels.

     The effect on the pre- and post-capillary resistances
R_a and R_v of vasoactive substances injected into the circulation can be observed indirectly by our method from the trans-

ient response of the arterial density. After the injection

of vasoconstricting drugs we found a transient decrease of

the density as shown in fig. 62 (Kenner et al. 1977 b).

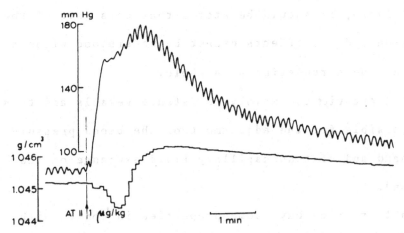

Fig. 62 Blood pressure (upper tracing) and arterial blood
        density in a cat after injection of 1 μg angioten-
        sin II (Kenner et al. 1977 b)

This decrease of the density indicates a relatively more

marked constriction of the precapillary resistance which

reduces the capillary pressure and thus leads to an influx

of interstitial fluid. Following this transient the densi-

ty tends to rise with rising pressure. We also observed this

parallelity of density and pressure during experiments in

dogs and during measurements in humans during hemodialysis

(Kenner et al. 1977 b). Particularly marked is the entry in-

to the blood of interstitial fluid immediately after blee-

ding. - Here a very fast reduction of blood density is a ty-

pical response (Kenner et al. 1977).

6. SUMMARY

In this article I have tried to present an overlook over the interesting problems related to blood flow in small arteries. Again, it should be stated that this part of the circulation and its effects cannot be understood without looking at the circulation as a whole.

Small arteries are mainly resistance vessels and thus are responsible for the adjustment of the blood pressure on one hand and of transcapillary fluid exchange on the other hand.

Small arteries have some properties in common with large vessels, and some properties in common with capillaries, especially as far as flow of erythrocytes and plasma is concerned.

My viewpoint in presenting these topics is certainly a personal one and I hope the reader will excuse the emphasis on our own work.

LITERATURE AND REFERENCES

Anliker, M., Histand, M.B. and Ogden, E. (1968), Dispersion and attenuation of small artificial pressure waves in the canine aorta, Circulat.Res.23, 639.

Attinger,E.O. and Attinger, F. (1973), Frequency dynamics of the peripheral vascular blood flow, Ann. Rev.Biophys. Bioengineering. 2, 7.

Basar, E. (1981) , Vasculature and circulation, Elsevier/ North Holland , Amsterdam-N.Y.-Oxford.

Basar, E., Basar-Eroglu, C., Demir, N., Tümer, N. and Weiss, C. (1982). The overall myogenic coordination in circulatory dynamics, pp. 509, from Kenner et al.(1.c.)

Bauer,R.D., Busse, R. (1979), The arterial system, dynamics, control theory and regulation. Springer, Berlin-Heidelberg.

Bauer, R.D.,Busse,R., Schabert,A. and Wetterer,E. (1982), The role of elastic and viscous wall properties in the mechanics of elastic and muscular arteries, pp.373,from Kenner et al. (1.c.)

Benninghoff, A., (1930), Blutgefäße und Herz. from: Handbuch der mikroskopischen Anatomie, Bd. VI/1 pp.1,Berlin.

Bergel, D.H. (1961), The dynamic elastic properties of the arterial wall. J. Physiol. (London) 156 , 458.

Broemser,Ph. (1932) , Beitrag zur Windkesseltheorie des Kreislaufs. Zeitschr. Biol. 93, 149.

Broemser, Ph. and Ranke, F. (1930),Über die Messung des
    Schlagvolumens des Herzens auf unblutigem Weg. Zeitschr.
    Biol. 90, 467.

Burton, A.C.(1951), On the physical equilibrium of small
    blood vessels. Amer.J.Physiol. 164, 319.

Burton, A.C. (1962), Physical principles of circulatory phe-
    nomena: the physical equilibrium of heart and blood
    vessels, from Handbook of physiology. Sect. 2, Circula-
    tion ,Vol. I, pp. 85 , Washington. D.C.

Busse,R.,Bauer,R.D., Burger,W.,Sturm,K. and Schabert,A.
    (1982), Correlation between amplitude and frequency
    of spontaneous rhythmic contractions and the mean circum-
    ferential wall stress of a small muscular artery, from
    Kenner et al. pp.363 (1.c.)

Busse,R., Sturm,K., Schabert,A. and Bauer,R.D. (1982 b),
    The contribution of the parallel and series elastic
    components to the dynamic properties of the rat tail
    artery under two different smooth muscle tones. Pflügers
    Arch. 393, 328.

Caro,C.G.,Pedley,T.J.,Schoter,R.C. and W.A.Seed,(1978),
    The mechanics of the circulation. Oxford University Press
    N.Y.-Toronto.

Cox,R.H. (1982), Determination of the mechanical properties
    of the contractile system in arterial smooth muscle
    models. pp.317, from Kenner et al. (1.c.)

Dobrin,P.B.,(1978), Mechanical properties of arteries.

Physiol. Rev. 58, 397.

Dujardin,J.P.L. and Scott,D.L. (1982), The dynamic arterial

pressure flow relationship and total arterial compliance

in spontaneously hypertensive and normal rats. pp. 199,

from Kenner et al. (l.c.)

Folkow,B. and Neil,E. (1971), Circulation, Oxford University

Press, N.Y.-London-Toronto.

Frank,O. (1899), Die Grundform des arteriellen Pulses.

Zeitschr. Biol. 37, 483.

Frank,O. (1906), Die Analyse endlicher Dehnungen und die

Elastizität des Kautschuks. Ann. Physik. 21, 602.

Frank,O. (1920), Die Elastizität der Blutgefäße. Zeitschr.

Biol. 71,255.

Fung, Y.C. (1981), Biomechanics, mechanical properties of

living tissues. Springerverlag, N.Y.-Heidelberg-Berlin.

Gessner,U. (1981) personal communication.

Gow.,B.S. (1980), Circulatory correlates: vascular impedan-

ce, resistance , and capacity. pp 353, from Handbook of

Physiology. Vol.II Vascular smooth muscle. Washington D.C.

Green,H.D. (1944), Circulation: Physical principles, from

O.Glasser (ed.) Medical Physics. Year Book Publ.,Chicago.

Green, H.D., Lewis,R.N.,Nickerson,N.D. and Heller,L.(1944),

Blood flow, peripheral resistance and vascular tonus,

with observations on relationship between blood flow and

cutaneous temperatures. Amer.J.Physiol. 141,518.

Gross, J.F.,(1977), The significance of pulsatile micro-
    hemodynamics. pp. 365, from G.Kaley and B.M.Altura (eds.)
    Microcirculation Vol.1, University Park Press, Baltimore.

Gross,J.F. and Popel,A.S. (1980), Mathematics of microcircu-
    lation phenomena. Raven Press, N.Y.

Guyton,A.C. and Cowley,A.W.,(1976), Cardiovascular Physiolo-
    gy II., University Park Press , Baltimore.

Hatakeyama, I.,(1982) , Hydrodynamic amplification in blood
    vessels and cardiovascular dynamics. pp.181, from Kenner
    et al. (l.c.)

Hudetz,A.G. and Monos,E.,(1982), A structural model for non-
    linear anisotropic behaviour of the arterial wall.
    pp. 337. from Kenner et al. (l.c.)

Iberall, A.S.,(1967) ,Anatomy and steady flow characteris-
    tics of the arterial system with an introduction to its
    pulsatile characteristics. Math. Biosci. 1,375.

Intaglietta, M., (1977), Transcapillary exchange of fluid
    in single microvessels. pp. 197 , from Kaley and Altura
    (l.c.)

Kaley,G. and Altura,B.M. (eds.) (1977). Mircocirculation
    Vol.I., University Park Press, Baltimore.

Kenner,T. (1967), Neue Gesichtspunkte und Experimente
    zur Beschreibung und Messung der Arterienelastizität.
    Archiv.Kreislaufforschung.54,68.

Kenner,T. (1971), Dynamic control of flow and pressure in
    the circulation. Kybernetik, 9,215.

Kenner,T. (1972), Flow and pressure in the arteries. from
    Fung,Y.C. et al. (eds.) , Biomechanics, its foundations
    and objectives. Prentice Hall. Englewood Cliffs N.J.

Kenner,T. (1974) , Beziehungen zwischen Dynamik und Re-
    gulation des Arteriensystems. Verh. Dtsch. Ges. Kreis-
    laufforschung. 40 , 41.

Kenner,T. (1975) , The central arterial pulses. Pflügers
    Arch. 353, 67.

Kenner, T. (1978) , Models of the arterial system. from
    R.D.Bauer and R.Busse (.eds.). The arterial system. pp.80,
    Springerverlag, Berlin-Heidelberg.

Kenner,T. (1979) , Physical and mathematical modeling in
    cardiovascular systems. pp.41,from N.H.C.Hwang et al.
    Quantitative cardiovascular studies. University Park
    Press.

Kenner,T. and Bergmann H. (1975), Frequency dynamics of
    arterial autoregulation. Pflügers Arch. 356,169.

Kenner,T., Busse,R. and Hinghofer-Szalkay,H. (1982) ,
    Cardiovascular system dynamics - models and measurements,
    Plenum Press. N.Y-London.

Kenner,T., Hinghofer-Szalkay,H., Leopold,H. and Pogglitsch,
    H. (1977 b). The relation between the density of blood
    and the arterial blood pressure in animal experiments
    and in patients during hemodialysis. Zeitschr. Kardiol.
    66, 399.

Kenner,T., Hinghofer-Szalkay,H.,Moser,M and Leopold,H.(1982)
The application of the continuous recording of blood density for hemodynamic measurements. pp. 431, from Kenner
et al. (l.c.)

Kenner,T., Moser,M. and Hinghofer-Szalkay,H.(1980) , Determination of cardiac output and transcapillary fluid exchange by continuous recording of blood density. Basic Res.
Cardiol. 75, 501.

Kenner,T. and Ono,K. (1971) , Reciprocal autoregulation of
blood flow and blood pressure. Experientia 27,528.

Kenner,T. and Ono,K, (1971) , The low frequency input impedance of the renal artery. Pflügers Arch. 324, 155.

Kenner,T. and Ono,K. (1972) , Humoral autoregulation of
blood flow and blood pressure. Experientia 28, 528.

Kenner, T. and Ono,K. (1972) , Analysis of slow autooscillations of arterial flow. Pflügers Arch. 331, 347.

Kenner,T. and Ono,K. (1972) , Interaction between circulatory control and drug-induced reactions. Pflügers Arch.
331,335.

Kenner,T., Ueda,M, Huntsman,L. and Attinger,E.O. (1968) ,
Effects of local and general hypoxia on iliac flow.
Angiology 5,345.

Kenner,T., van Zwieten,P.A. (1982), Use of hemodynamic
analysis for the interpretation of the mode of action
of vasoactive drugs. from Kenner et al. (l.c.)

Klitzman,B. and Duling, B.R. (1979), Microvascular hemato-

crit and red blood cell flow in resting and contracting
striated muscle. Amer.J.Physiol. 273, H 481.

Lee,J.S. (1980) , Micro-macroscopic scaling. pp.159 from
Gross and Popel (l.c.)

Lee, J.S. and Kenner,T. (1982) , Microvascular dynamics.
from Kenner et al. pp. 413, (l.c.)

Lee, J.S. and Nellis,S. (1974) , Modeling studies on the
distribution of flow and volume in the microcirculation
of cat mesentery. Ann. Biomed. Eng. 2, 206.

Lefèvre, J. (1982) , Teleonomical representation of the
pulmonary arterial bed of the dog by a fractal tree.
pp. 137, from Kenner et al. (l.c.)

Leopold, H., Jellinek,R. and Tilz,G. (1977) , The application of the mechanical oscillator technique for the determination of the density of physiological fluids. Biomed.
Technik. 22, 231.

Lindner,A. and Ronniger,R. (1955) , Zur Darstellung der
Beziehungen zwischen zentralen und peripheren Pulsen als
Ortskurven. Arch. Kreislaufforschung. 22, 72.

Lipowsky,H.H., Usami,S. and Chien,S. (1980) , In vivo measurements of "apparent viscosity" and microvascular hematocrit in the mesentery of the cat. Microvasc. Res.19,
297.

Lipowsky,H.H. and Zweifach,B.W. (1974), Network analysis
of microcirculation of cat mesentery. Microvasc. Res.
7, 73.

Mayrovitz,H.N.,Wiedeman,M.P. and Noordergraaf,A. (1976) ,
    Analytical characterization of microvascular resistance
    distribution. Bull. Math. Biphys. 38,71.

McDonald,D.A. (1974) , Blood flow in arteries, 2 nd ed.
    Edward Arnold, London.

Metzger,H. (1973) , Geometric considerations in modeling
    oxygen transport processes in tissue. Advances Exp.Biol.
    Med. 376,661.

Monos,E. and Kovach, A.G.B. (1982) , Biomechanics of isola-
    ted canine splenic artery. pp.327, from Kenner et al.(1.c.)

Newman,D.L. and Greenwald,S.E. (1982) The effect of smooth
    muscle activity on the static and dynamic properties of
    the rabbit carotid artery. pp. 393 , from Kenner et al.
    (1.c.)

Noordergraaf,A. (1978) , Circulatory systems dynamics.
    Academic Press, N.Y.

O´Rourke, M.F. (1982) , Vascular impedance in studies of
    arterial and cardiac function. Physiol. Rev. 62, 570.

O´Rourke, M.F. (1982) , Vascular impedance - a call for
    standardization. pp. 175, from Kenner et al. (1.c.)

Patel, D.J.,Austen,W.G. and Greenfield,J.C. (1964) , Impe-
    dance of certain large blood vessels in man. Ann. N.Y.
    academy Sci. 115,1129.

Patel,D.J.,Freitas,F.M.,Greenfield,J.C. and Fry,D.L.(1963),
    Relationship of radius to pressure along the aorta in
    living dogs. J.Appl. Physiol. 18,1111.

Patel,D.J.,Vaishnav,R.N. and Atabek,H.B. (1979) , Local
mechanical properties of the vascular intima and adjacent flow fields. p. 215,from Hwang et al. Quantitative
cardiovascular studies. University Park Press.

Pollak,G.H.,Reddy,R.V. and Noordergraaf,A. (1968), Input
impedance, wave travel and reflections in the pulmonary
arterial tree. Studies using an electric analog. IEEE
transact. BME. 15,151.

Popel,A.S. (1980) Mathematical modeling of convective and
diffusive transport in the microcirculation. pp.63,from
Gross and Popel (l.c.)

Rhodin, J.A.G. (1980), Architecture of the vessel wall.
pp. 1,from Handbook of Physiology,Vol.II, Vascular
smooth muscle, Washington D.C.

Ronniger,R.,(1954) , Über eine Methode der übersichtlichen
Darstellung hämodynamischer Zusammenhänge. Arch. Kreislaufforschung. 21,127.

Ronniger,R., (1955). Zur Theorie der physikalischen Schlag-
volumenbestimmung. Arch. Kreislaufforschung. 22,332.

Rosen,R. (1967), Optimality principles in biology.
Butterworths, London.

Rubenstein,H.J.,Kenner,T. and Ono,K. (1973), Pseudorandom
test technique for the characterization of local hemodynamic control. Pflügers Arch. 343,309.

Schimmler, W.,(1965), Untersuchungen zum Elastizitätspro-
blem der Aorta. Arch.Kreislaufforschung ,47,189.

Schleier,J. (1918). Der Energieverbrauch der Blutbahn.
    Pflügers Arch. 173,172.

Schmid-Schönbein,H. (1976) , Microrheology of erythrocytes,
    blood viscosity and the distribution of blood flow in
    the microcirculation , pp.1,from Guyton and Cowley (l.c.)

Sipkema,P. and Westerhof,N. (1982), Peripheral resistance
    and low frequency impedance of the femoral bed. pp.501,
    from Kenner et al. (l.c.)

Taylor,G. (1953), Dispersion of soluble matter in solvent
    flowing slowly through a tube. Proc.Royal Soc.Lond.
    Ser.A. 219,186.

Taylor,M.G., (1966), Use of random excitation and spectral
    analysis in the study of frequency dependent parameters
    of the cardiovascular system. Circulat. Res. 18,585.

Vadot,L. (1967), Mécanique du coeur et des arteres.
    L´expansion scientifique Franc., Paris.

Van Loon,P.,Klip,W. and Bradley,E.L. (1977), Length-force
    and volume-pressure relationship of arteries.
    Biorheology, 14,181.

Weizsäcker,H.W. and Pascale,K. (1977), Das Kraft-Dehnungs-
    verhalten von Rattenkarotiden in Längsrichtung bei ver-
    schiedenem Innendruck und seine modulmäßige Deutung.
    Basic Res.Cardiol. 72,619.

Weizsäcker,H.W. and Pascale,K.(1982), Anisotropic passive
    properties of blood vessel walls. pp.347, from Kenner
    et al. (l.c.)

Westerhof,N.,Bosman,F.,De Vries,C.J. and Noordergraaf,A.(1969)
J. Biomechanics 2,121.

Westerhof,N.,Sipkema,P.,Elzinga,G,Murgo,J.P. and Giolma,J.P.,
(1979), Arterial impedance, pp. 111, from Hwang et al. ,
Quantitative cardiovascular studies, University Park Press.

Wetterer,E.,Bauer,R.D. and Busse.R. (1977) , Arterial dynamics. INSERM-Euromech 91, Cardiovascular and pulmonary
dynamics,Vol.71 ,pp.17.

Wetterer,E. and Kenner,T. (1968), Grundlagen der Dynamik
des Arterienpulses. Springerverlag, Berlin-N.Y.Heidel-
berg.

Wetterer,E. and Pieper,H. (1953 a), Über die Gesamtelasti-
zität des arteriellen Windkessels und ein experimentelles
Verfahren zu ihrer Bestimmung am lebenden Tier. Zeitschr.
Biol. 106,23.

Wetterer,E.and Pieper,H. (1953 b), Messungen am Arterien-
system in vivo während erzwungener periodischer Volumen-
schwankungen. Verh.Dtsch.Ges.Kreislaufforschung 19,259.

Wezler,K. and Sinn,W.,(1953), Das Strömungsgesetz des Blut-
kreislaufs. Ed.Cantor,Aulendorff i. Württemberg.

Whitmore.R.L. (1968) ,Rheology of the Circulation. Pergamon
Press, Oxford-London-N.Y.

Wiedemann.M.P. (1962), Lengths and diameters of peripheral
arterial vessels in living animals. Circulat.Res.10, 686.

Wiedeman, M.P. (1963), Dimensions of blood vessels from distributing artery to collecting vein. Circulat.Res. 12,375.

Witzig,K. (1914), Über erzwungene Wellenbewegungen zäher,

inkompressibler Flüssigkeiten in elastischen Röhren.

Dissertation,Bern.

Womersley,J.R. (1957), An elastic tube theory of pulse

transmission and oscillatory flow in mammalian arteries.

WADC Report, TR 65-614.

CHAPTER V

FLOW IN LARGE ARTERIES

Czeslaw M. Rodkiewicz
Faculty of Engineering
The University of Alberta
Edmonton, Alberta, Canada

1. INTRODUCTION

1.1 Introduction

For centuries, the world within himself fascinated man as much as
his near and distant environment. In particular the cardiovascular system was the object of attention of scientific observers like Aristotle
and Leonardo da Vinci. However, the concept of the Circulation of the
Blood was clearly presented by W. Harvey in 1628, in his famous De Motu
Cordis et Sanguinis in Animalibus. The evidence provided was almost complete and reached into the present day understanding, except that Harvey
could not see the passage of blood from the peripheral arteries to the
veins. He speculated that there must be "pores" at these locations.
These "pores" were in 1661 identified by Malpighi as the capillaries
(K.D. Keele, 1978). Later in 1733 Stephen Hales, the Vicar of Teddington,

published the first measurements of arterial blood pressure.

The circulatory system, with the heart as the driving force of the double-action pump type, serves to transport and deliver to the tissues those substances which are essential for maintenance of function and to remove the by-products of metabolism. The vessels that arise from the two ventricles and serve as the delivery roots constitute the pulmonary and systemic arteries. The systemic arteries are considered to be the primary distributing conduits, however, the smaller arteries, and in particular the arterioles, maintain the blood pressure and are instrumental in regulation of flow rate to the respective tissues. Elaborate drawing of the maze of arteries running through the body can be found in any anatomy text book. Originating at the aorta these distributing conduits subdivide the flowing blood into many streams by the Y type bifurcations, most of which are non-symmetric. The aorta resembles an inverted U with one end leading out of the heart and the other end leading down into the abdomen. Normally it has three channels branching from the top of the curve to the head and upper body.

The heart produces a periodic or pulsatile flow on the arterial side of the circulatory system. The amplitude of the flow pulse is largest in the aorta and becomes gradually smaller as the system branches. The arterial vessels are subjected to higher pressure and pressure variation, and they are thicker and contain more elastin than the venous system. Despite the extra strength and elasticity of the arterial walls, it seems likely that a system under the continual wear and tear of a pulsatile flow would be subject to many disorders. Such is the case; arterial disease is a very great problem.

Among diseases of the arterial tree atherosclerosis is the most common and the most important. Under the general terms of atherosclerosis are included several types of tissue changes. Despite the abundance of relevant information atherosclerosis is little understood except in rather broad terms. Its apparently complex pathology so far defies precise explanation. There are few theses explaining the etiology of the atherosclerotic formations. These have been discussed, for example by Constantinides (1965).

One of the basic features of atherosclerosis is that it occurs predominantly at specific sites of the arterial net. It seems appropriate, therefore, to consider its association with the blood flow phenomena, which is implicated here in the etiology. The regions most susceptible to the atherosclerotic lesions are portions of larger arteries such as bends and junctions of various geometry. Although it is understood that hemodynamic forces have something to do with the lesion sites, the exact mechanism of inflicting these lesions has not yet been elucidated.

Considering the fluid-flow point of view, one cannot fail to notice that the above mentioned sites of lesions are also locations of developing stagnation and separation regions. Some correlation between the occurrence of the atherosclerotic plaque and the angle of bending and branching have been noted by Schneck and Gutstein (1966). The possible correlation between turbulent flow, thrombosis, and arterial lesions has been also pointed out by Mitchell and Schwartz (1965). Rodkiewicz (1975) postulated that the atherosclerotic formations of the aortic arch commence and develop at the locations where there are developed or developing sep-

aration and stagnation regions. This could be extended to other geometrical configurations, for examples bifurcations.

1.2 Newtonian Behavior of Blood

In the case of fluid motion, such as the flow in the cardiovascular
system, the influence of viscosity is such that, under the assumption of
the no-slip condition at the wall, the frictional forces retard the motion
of the fluid in a layer near the wall. In this layer, the boundary layer,
the velocity of the fluid increases from zero at the wall to its full
value at a distance from the wall. These velocity gradients across the
fluid stream give rise to the frictional shearing stresses (particularly
important at the wall), which when integrated yield the viscous drag on
the wall. To overcome this drag the heart has to expend associated energy.
According to Newton's law of friction the shear stress at the wall is proportional to the product of the absolute viscosity (viscosity describes
the resistance of a fluid to shear when the fluid is subjected to a tan-
gential stress) and the rate of deformation. Though blood is recognized
to be a non-Newtonian fluid it may be considered a Newtonian fluid when
flowing through conduits of a larger diameter. Consequently, for such
cases Newtonian fluids have been used as transport media when modelling
the blood flow. Weiting (1968), for example, found that a 36.7% glycerol-
aqueous solution was a good hydraulic analog for blood.

1.3 Separation and Stagnation Regions

In some cases the decelerated fluid particles do not follow the directions suggested by the containing walls. The boundary layer, in the

downstream direction, may increase significantly and the flow next to the
wall may become reversed. In such cases the decelerated fluid stream becomes separated from the wall as indicated in Fig. 1.3.1. This is associated with additional energy losses. The boundary layer separation

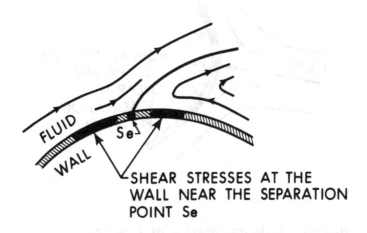

SHEAR STRESSES AT THE
WALL NEAR THE SEPARATION
POINT Se

Fig. 1.3.1 Diagrammatic Representation of Flow Near a Point of
            Boundary Layer Separation.

exists in regions with an adverse pressure gradient and the likelihood of
its occurrence increases at flow dividing elements, or in flows around the
curved walls. At the point of separation, Se, the shear stress is zero,
but on each side of it the shear stresses are finite and act in the direction toward the point, or line segment, Se (converging). A prerequisite
for a separation to appear is a prior development of a low shear stress
region.

Another pertinent case, so called free stagnation flow, is shown in

Fig. 1.3.2. Fluid impinges on the wall creating a stagnation point, or

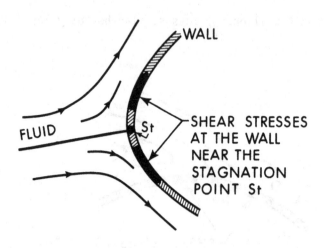

Fig. 1.3.2 Diagrammatic Representation of Flow Near a
           Stagnation Point.

line segment, St. Along the streamline which leads to the stagnation
point the pressure increases in the direction of flow. On each side of
that point the shear stresses are finite and act in the direction away
from the point St (diverging).

1.4 Flow Governing Parameters

     According to Kuchar and Ostrach (1965), and also Kuchar and Scala
(1968), the flow in a segment of arterial tree of a fixed geometry is described by three dimensionless parameters: Reynolds number, frequency
parameter, and the amplitude parameter. These are, respectively,

$$\text{Re} = UD/\nu, \quad \alpha = a(n/\nu)^{1/2}, \quad \lambda = U'/U \qquad (1.4.1, 1.4.2, 1.4.3)$$

where D = unstressed characteristic diameter, a = unstressed character-

istic radius, U = characteristic longitudinal velocity, U$'$ = peak to peak

amplitude of the fluctuating component of longitudinal velocity, ν =

kinematic fluid viscosity, and n = pulse rate.

1.5 Arterial Passage Classification

It is proposed to classify the large arteries into four major geo-

metrical types each with a descriptive set of dimensionless parameters.

The first class is considered to be straight arteries with the possibility

of a converging or diverging nature. The second class introduces curva-

ture to the straight artery, again with the possible complication of con-

vergence or divergence. The third class is composed of tubes with bi-

furcations or branches, and includes the possibilities that the tubes may

be curved, and converging or diverging. The fourth class are the most

complicated systems such as the aortic arch which combines all the aspects

of the first three groups.

2. BASIC EQUATIONS

2.1 The Governing Equations

The differential form of the law of conservation of mass can be

written in the form

$$\frac{\partial \rho}{\partial t} + \nabla \cdot (\rho \vec{V}) = 0 \qquad\qquad (2.1.1)$$

where ρ is density, t is time, and $\vec{V}$ is velocity vector. Blood may be

considered as an incompressible fluid and, therefore, the above equation

may be reduced as follows

$$\nabla \cdot \vec{V} = 0 \qquad (2.1.2)$$

The differential form of the principle of momentum, for constant fluid properties and a Newtonian fluid, may be written as

$$\frac{D\vec{V}}{Dt} = \vec{f} - \frac{1}{\rho} \nabla p + \frac{\mu}{\rho} \nabla^2 \vec{V} \qquad (2.1.3)$$

where $\vec{f}$ denotes the external force, p is pressure, μ is absolute viscosity, and

$$\nabla^2 \vec{V} = \nabla(\nabla \cdot \vec{V}) - \nabla \times (\nabla \times \vec{V}) \qquad (2.1.4)$$

2.2 The Equations of Motion in Rectangular Coordinates

Equations of motion (2.1.2) and (2.1.3), for later reference, may now be written in rectangular coordinates. If x,y,z denote a three-dimensional system of coordinates, and V_x, V_y, V_z denote the velocity components in the corresponding directions, then the following system of equations is obtained:

$$\frac{\partial V_x}{\partial x} + \frac{\partial V_y}{\partial y} + \frac{\partial V_z}{\partial z} = 0 \qquad (2.2.1)$$

$$\frac{\partial V_x}{\partial t} + V_x \frac{\partial V_x}{\partial x} + V_y \frac{\partial V_x}{\partial y} + V_z \frac{\partial V_x}{\partial z} = g_x - \frac{1}{\rho} \frac{\partial p}{\partial x}$$

$$\qquad (2.2.2)$$

$$+ \nu \left(\frac{\partial^2 V_x}{\partial x^2} + \frac{\partial^2 V_x}{\partial y^2} + \frac{\partial^2 V_x}{\partial z^2} \right)$$

$$\frac{\partial V_y}{\partial t} + V_x \frac{\partial V_y}{\partial x} + V_y \frac{\partial V_y}{\partial y} + V_z \frac{\partial V_y}{\partial z} = g_y - \frac{1}{\rho} \frac{\partial p}{\partial y}$$

$$+ \nu \left(\frac{\partial^2 V_y}{\partial x^2} + \frac{\partial^2 V_y}{\partial y^2} + \frac{\partial^2 V_y}{\partial z^2} \right)$$

(2.2.3)

$$\frac{\partial V_z}{\partial t} + V_x \frac{\partial V_z}{\partial x} + V_y \frac{\partial V_z}{\partial y} + V_z \frac{\partial V_z}{\partial z} = g_z - \frac{1}{\rho} \frac{\partial p}{\partial z}$$

$$+ \nu \left(\frac{\partial^2 V_z}{\partial x^2} + \frac{\partial^2 V_z}{\partial y^2} + \frac{\partial^2 V_z}{\partial z^2} \right)$$

(2.2.4)

where ν is kinematic viscosity ($\nu = \mu/\rho$) and various g's represent acceleration components of the gravitational field.

2.3 The Equations of Motion in Cylindrical Coordinates

Equations of motion (2.1.2) and (2.1.3), for the passages of circular cross-section, may conveniently be written in cylindrical coordinates. If r, θ, z denote the radial, azimuthal, and axial coordinates respectively, of a three-dimensional system of coordinates, and V_r, V_θ, V_z denote the velocity components in the corresponding directions, then the following system of equations is obtained:

$$\frac{1}{r} \frac{\partial}{\partial r} (rV_r) + \frac{1}{r} \frac{\partial V_\theta}{\partial \theta} + \frac{\partial V_z}{\partial z} = 0$$

(2.3.1)

$$\frac{\partial V_r}{\partial t} + V_r \frac{\partial V_r}{\partial r} + \frac{V_\theta}{r} \frac{\partial V_r}{\partial \theta} - \frac{V_\theta^2}{r} + V_z \frac{\partial V_r}{\partial z} = g_r - \frac{1}{\rho} \frac{\partial p}{\partial r}$$

$$(2.3.2)$$

$$+ \nu \left[\frac{\partial}{\partial r} \left(\frac{1}{r} \frac{\partial}{\partial r} (rV_r) \right) + \frac{1}{r^2} \frac{\partial^2 V_r}{\partial \theta^2} - \frac{2}{r^2} \frac{\partial V_\theta}{\partial \theta} + \frac{\partial^2 V_r}{\partial z^2} \right]$$

$$\frac{\partial V_\theta}{\partial t} + V_r \frac{\partial V_\theta}{\partial r} + \frac{V_\theta}{r} \frac{\partial V_\theta}{\partial \theta} + \frac{V_r V_\theta}{r} + V_z \frac{\partial V_\theta}{\partial z} = g_\theta - \frac{1}{\rho} \frac{\partial p}{\partial \theta}$$

$$(2.3.3)$$

$$+ \nu \left[\frac{\partial}{\partial r} \left(\frac{1}{r} \frac{\partial}{\partial r} (rV_\theta) \right) + \frac{1}{r^2} \frac{\partial^2 V_\theta}{\partial \theta^2} + \frac{2}{r^2} \frac{\partial V_r}{\partial \theta} + \frac{\partial^2 V_\theta}{\partial z^2} \right]$$

$$\frac{\partial V_z}{\partial t} + V_r \frac{\partial V_z}{\partial r} + \frac{V_\theta}{r} \frac{\partial V_z}{\partial \theta} + V_z \frac{\partial V_z}{\partial z} = g_z - \frac{1}{\rho} \frac{\partial p}{\partial z}$$

$$(2.3.4)$$

$$+ \nu \left[\frac{1}{r} \frac{\partial}{\partial r} \left(r \frac{\partial V_z}{\partial r} \right) + \frac{1}{r^2} \frac{\partial^2 V_z}{\partial \theta^2} + \frac{\partial^2 V_z}{\partial z^2} \right]$$

In the above, the term V_θ^2/r when multiplied by ρ, yields the centrifugal force. Similarly, expression $\rho \, V_r V_\theta/r$ is the Coriolis force.

2.4 The Shear Stresses

The components of the local shear stress associated with Equations (2.2.1) through (2.2.4) may be written in the following way.

$$\tau_{xy} = \tau_{yx} = \mu \left(\frac{\partial V_x}{\partial y} + \frac{\partial V_y}{\partial x} \right)$$

$$(2.4.1a)$$

$$\tau_{yz} = \tau_{zy} = \mu \left(\frac{\partial V_y}{\partial z} + \frac{\partial V_z}{\partial y} \right)$$

$$(2.4.1b)$$

$$\tau_{zx} = \tau_{xz} = \mu \left(\frac{\partial V_z}{\partial x} + \frac{\partial V_x}{\partial z} \right) \tag{2.4.1c}$$

Similarly the local shear stress components associated with Equations (2.3.1) through (2.3.4) are as follows.

$$\tau_{r\theta} = \tau_{\theta r} = \mu \left[r \frac{\partial}{\partial r} \left(\frac{V_\theta}{r} \right) + \frac{1}{r} \frac{\partial V_r}{\partial \theta} \right] \tag{2.4.2a}$$

$$\tau_{\theta z} = \tau_{z\theta} = \mu \left(\frac{\partial V_\theta}{\partial z} + \frac{1}{r} \frac{\partial V_z}{\partial \theta} \right) \tag{2.4.2b}$$

$$\tau_{zr} = \tau_{rz} = \mu \left(\frac{\partial V_z}{\partial r} + \frac{\partial V_r}{\partial z} \right) \tag{2.4.2c}$$

A double-index scheme has been utilized. The first subscript denotes the direction of the normal to the plane containing the stress, while the second subscript indicates the direction of the stress itself. For example τ_{zx} is the value of the shear stress acting in a plane whose normal is parallel to the z direction, while the stress itself is parallel to the x direction.

3. FLOW IN STRAIGHT PASSAGES

3.1 Introduction

In the case of a simple straight artery of constant diameter the velocity and pressure distributions would be given by the Poiseuille relationship provided the flow is fully developed. However, in the arteries in which blood may be assumed to behave like a Newtonian fluid, this condition most likely does not exist. Since the flow rate must be the same for every passage section the decrease in the volume of flow near the wall, which is due to friction, must be compensated by a corresponding increase near the axis. Consequently, the boundary layer is formed under the influence of an accelerating external stream. Such a steady state flow will never separate from a regular wall and the shear stress at the wall will tend to transform asymptotically to the value which would exist if Poiseuille flow could be established. In the pulsatile case adverse pressure gradients may be present and, depending on the fluctuation frequency and amplitude, reversed flow at the wall could appear. The associated minimum wall shear stress would be equal to zero and the wall drag would keep changing its direction.

For the steady state flow in the converging and diverging passages the incompressible fluid accelerates and decelerates, respectively. In the latter case, when channel divergence is sufficiently significant, the boundary layer may not be able to cope with the adverse pressure gradient and the flow may separate from the wall. In a closed circuit it will reattach further downstream, and will enclose a separation region. At the locus of separation points the shear stress at the wall is equal to zero. The wall shear stresses within the separation region, which generates vortices shed into the main stream (not to be identified with the onset of turbulence), are in the direction opposite to the direction of the swifter main stream.

In general the separation region should not vanish in the pulsatile flow. It should, however, change its shape and shift back and forth in accord with the pulsation frequency.

At this point it is of interest to mention a study done by M.J. Martin (1960) which illustrates the presence of atherosclerotic plaque in the carotid sinus. This is where channel divergence has been observed.

3.2 Flow in a Rigid Tube of Elliptic Cross-Section

Consider a steady and fully developed flow in which the body forces may be neglected. Let the z-direction be the direction of flow. For such a case Equations (2.2.1) through (2.2.4) reduce to the form:

$$\frac{\partial u}{\partial x} + \frac{\partial v}{\partial y} = 0 \tag{3.2.1}$$

$$u\frac{\partial u}{\partial x} + v\frac{\partial u}{\partial y} = -\frac{1}{\rho}\frac{\partial p}{\partial x} + \nu\left(\frac{\partial^2 u}{\partial x^2} + \frac{\partial^2 u}{\partial y^2}\right) \tag{3.2.2}$$

$$u\frac{\partial v}{\partial x} + v\frac{\partial v}{\partial y} = -\frac{1}{\rho}\frac{\partial p}{\partial y} + \nu\left(\frac{\partial^2 v}{\partial x^2} + \frac{\partial^2 v}{\partial y^2}\right) \tag{3.2.3}$$

$$u\frac{\partial w}{\partial x} + v\frac{\partial w}{\partial y} = -\frac{1}{\rho}\frac{\partial p}{\partial z} + \nu\left(\frac{\partial^2 w}{\partial x^2} + \frac{\partial^2 w}{\partial y^2}\right) \tag{3.2.4}$$

where for V_x, V_y, V_z we have written u, v, w respectively.

Differentiating expression (3.2.2) with respect to y, and expression (3.2.3) with respect to x and eliminating the pressure terms by subtracting one of these equation from the other, one obtains

$$u\frac{\partial}{\partial x}\left(\frac{\partial u}{\partial y} - \frac{\partial v}{\partial x}\right) + v\frac{\partial}{\partial y}\left(\frac{\partial u}{\partial y} - \frac{\partial v}{\partial x}\right) = \nu\left(\frac{\partial^2}{\partial x^2} + \frac{\partial^2}{\partial y^2}\right)\left(\frac{\partial u}{\partial y} - \frac{\partial v}{\partial x}\right) \tag{3.2.5}$$

Let us now introduce the stream function ψ (x,y), so that Equation (3.2.1) is satisfied automatically by the following expressions

$$u = \frac{\partial \psi}{\partial y}, \quad v = - \frac{\partial \psi}{\partial x} \qquad\qquad (3.2.6)$$

Now, with the aid of (3.2.6), Equation (3.2.5) can be written in terms of the stream function, namely

$$u \frac{\partial}{\partial x} \nabla^2 \psi + v \frac{\partial}{\partial y} \nabla^2 \psi = \nu \nabla^2 \nabla^2 \psi \qquad\qquad (3.2.7)$$

The u and v velocity components are equal to zero on the boundary. Therefore, according to expressions (3.2.6) function ψ is equal to a constant on the boundary. But function ψ = constant also satisfies Equation (3.2.7) and, consequently, function ψ = constant is the solution. This in turn means that the u and v velocity components are equal to zero at all points of the flow cross-section. If so, Equations (3.2.2) and (3.2.3) indicate that $\partial p/\partial x = \partial p/\partial y = 0$, and Equation (3.2.4) assumes the following form

$$\frac{\partial^2 w}{\partial x^2} + \frac{\partial^2 w}{\partial y^2} = \frac{1}{\mu} \frac{dp}{dz} = \text{CONSTANT} \qquad\qquad (3.2.8)$$

Equation (3.2.8) yields

$$w = \frac{1}{2\,\mu\,(1 + \lambda^2)} \, (\lambda^2 x^2 + y^2 - \lambda^2 a^2) \frac{dp}{dz} \qquad\qquad (3.2.9)$$

where $\lambda = b/a$, and 2a and 2b are the major and minor axes of a cross-section, respectively. Also

$$w_{mean} = \frac{1}{2} w_{max} = - \frac{\lambda^2 a^2}{4 \mu (1 + \lambda^2)} \frac{dp}{dz} \qquad (3.2.10)$$

In addition, using the following definition for the friction

coefficient

$$f = \frac{\tau_{wall}}{\frac{1}{2} \rho w_{mean}^2} , \qquad \tau_{wall} = \frac{(p_1 - p_2) ab}{(a + b) k\ell} \qquad (3.2.11, 3.2.12)$$

we obtain (ℓ is length of tube):

$$f = \frac{8}{Re} \qquad (3.2.13)$$

where

$$Re = \frac{w_{mean} L}{\nu} , \quad L = K \frac{\lambda a (1 + \lambda)}{(1 + \lambda^2)} ,$$

$$K = 1 + m^2/4 + m^4/64 + m^6/256 + \dots , \quad m = \frac{1 - \lambda}{1 + \lambda}$$

3.3 The Hagen-Poiseuille Flow

The solution for flow through a tube of elliptic cross-section can

be specialized to the case of flow through a tube of circular cross-section.

Letting a = b, we obtain $\lambda = 1$, and from Equation (3.2.9) the velocity

component parallel to the axis becomes

$$w = \frac{1}{4\,\mu} (r^2 - a^2) \frac{dp}{dz} \qquad\qquad (3.3.1)$$

where $r^2 = x^2 + y^2$. For this case m = 0, K = 1, L = a, and the Reynolds
number based on the tube radius becomes Re = $w_{mean}a/\nu$. The velocity
over a cross-section is distributed in the form of a paraboloid of revolution. This type of laminar flow occurs as long as the tube Reynolds
number has a value which is less than the critical Reynolds number. For
large Reynolds numbers the flow may become turbulent and the present
analysis becomes invalid.

At this time it should be indicated that the generally adopted
practical definition of the friction factor for pipe flow calculations,
which differs from definition (3.2.11), is the Darcy-Weisbach equation,
namely

$$h = f \frac{\ell}{D} \frac{V^2}{2g} \qquad\qquad (3.3.2)$$

where h is the head loss (meter-newton per newton); ℓ is length of pipe;
D is pipe diameter; V is mean velocity; and g is the gravitational constant. In Equation (3.2.10) the pressure gradient dp/dz is a constant,
and when used in conjunction with Equation (3.3.2) yields

$$f = \frac{64}{Re} \qquad\qquad (3.3.3)$$

where Re = VD/ν.

3.4 Pulsating Flow in Rigid Circular Tube

Blood in the arterial system moves under the influence of the variable pressure gradients. In order to better understand the pulsatile blood flow some simplified analytical models have been developed. At this time we will involve ourselves with the solution of S. Uchida (1956), which is similar to that given by J.R. Womersley (1955). In the limit these solutions reduce to the steady state Hagen-Poiseuille equations provided in Section 3.3.

The associated simplifying assumptions were: the artery is regarded as a rigid tube of constant diameter; the velocity distribution across the flow section is independent of the coordinate along the direction of flow; the velocity is purely axial everywhere; there are no body forces; density is constant; viscosity is constant; the flow is laminar; and the fluid is a Newtonian fluid. It is most unlikely that all of these conditions are met in the arterial tree - yet the solutions serve a useful purpose. For such a flow, equations (2.3.1) through (2.3.4) reduce to a single equation used by Uchida (1956) and by Womersley (1955) as the starting point, namely

$$\frac{\partial w}{\partial t} = -\frac{1}{\rho}\frac{\partial p}{\partial z} + \nu \left(\frac{\partial^2 w}{\partial r^2} + \frac{1}{r}\frac{\partial w}{\partial r}\right) \tag{3.4.1}$$

where r is radius, z is the flow direction and w is velocity in z direction.

The pressure gradient in (3.4.1) becomes a function of time only and it was expressed by a Fourier series as follows

$$- \frac{1}{\rho} \frac{\partial p}{\partial z} = P_0 + \sum_{n=1}^{\infty} (P_{cn} \cos nt + P_{sn} \sin nt) \qquad (3.4.2)$$

where P_0 corresponds to the Hagen-Poiseuille flow pressure gradient, and P_{cn} and P_{sn} are constants representing the amplitudes of the time contributing terms. The corresponding expression for the velocity distribution was assumed to be

$$w = W_0 + \sum_{n=1}^{\infty} (W_{cn} \cos nt + W_{sn} \sin nt) \qquad (3.4.3)$$

where W_0 is the Hagen-Poiseuille velocity distribution given by expression (3.3.1), and W_{cn} and W_{sn} are constants representing the amplitudes of the time contributing terms.

Expressions (3.4.1) through (3.4.3) yield the dimensionless velocity distribution ($k = \sqrt{n/\nu}$; note that $ak = \alpha$) given by

$$\frac{w}{W} = 2 \left(1 - \frac{r^2}{a^2}\right) + \frac{S}{W}$$

$$= 2\left(1 - \frac{r^2}{a^2}\right) + \sum_{n=1}^{\infty} \frac{P_{cn}}{P_0} \left[\frac{8 B}{(ka)^2} \cos nt + \frac{8(1 - A)}{(ka)^2} \sin nt \right]$$

$$+ \sum_{n=1}^{\infty} \frac{P_{sn}}{P_0} \left[\frac{8 B}{(ka)^2} \sin nt - \frac{8(1 - A)}{(ka)^2} \cos nt \right] \qquad (3.4.4)$$

where

$$A = \frac{\text{ber } ka \text{ ber} kr + \text{bei} ka \text{ bei } kr}{\text{ber}^2 ka + \text{bei}^2 ka} \qquad (3.4.5a)$$

$$B = \frac{bei\ ka\ ber\ kr - berka\ bei\ kr}{ber^2 ka + bei^2 ka} \tag{3.4.5b}$$

and where W is the mean velocity of the W_0 distribution. The associated

pressure gradient becomes

$$-\frac{2a}{\frac{1}{2}\rho W^2}\frac{\partial p}{\partial z} = \frac{64}{Re}\left[1 + \sum_{n=1}^{\infty}\left(\frac{P_{cn}}{P_0}\cos nt + \frac{P_{sn}}{P_0}\sin nt\right)\right] \tag{3.4.6}$$

where Re = $2aW/\nu$.

In conclusion if a pressure gradient is presented as a Fourier

series, then the corresponding velocity can be computed from equation

(3.4.4). It will be noted that the periodic part of the velocity distri-

bution is governed by the parameter (ka).

Uchida (1956) also obtained velocity distributions for the limiting

cases of a very small and a very large magnitude of the parameter (ka).

In the latter case the motions near the center of the tube and near the

wall of the tube were discussed.

For the case when (ka) < < 1 expression (3.4.4) reduces to the

velocity distribution given by a paraboloid of revolution as is the case

for the steady Hagen-Poiseuille flow, namely

$$w = \frac{1}{4\nu}(a^2 - r^2)\left(-\frac{1}{\rho}\frac{\partial p}{\partial z}\right) \tag{3.4.7}$$

It is seen that the magnitude of velocity varies in phase with the pres-

sure gradient.

When the parameter (ka) > 10 the motion near the center of the tube was given as

$$\frac{w}{W} = 2 \left(1 - \frac{r^2}{a^2}\right) + \sum_{n=1}^{\infty} \frac{P_{cn}}{P_0} \frac{8}{(ka)^2} \cos\left(nt - \frac{\pi}{2}\right)$$

$$+ \sum_{n=1}^{\infty} \frac{P_{sn}}{P_0} \frac{8}{(ka)^2} \sin\left(nt - \frac{\pi}{2}\right) \tag{3.4.8}$$

and near the wall by

$$\frac{w}{W} = 2\left(1 - \frac{r^2}{a^2}\right)$$

$$+ \sum_{n=1}^{\infty} \frac{P_{cn}}{P_0} \frac{8}{(ka)^2} \left\{ \sin nt - \sqrt{\frac{a}{r}} \, e^{-k(a-r)/\sqrt{2}} \sin\left[nt - \frac{k}{\sqrt{2}}(a-r)\right]\right\}$$

$$+ \sum \frac{P_{sn}}{P_0} \frac{8}{(ka)^2} \left\{-\cos nt + \sqrt{\frac{a}{r}} \, e^{-k(a-r)\sqrt{2}} \cos\left[nt - \frac{k}{\sqrt{2}}(a-r)\right]\right\}$$

$$\tag{3.4.9}$$

Expression (3.4.8) indicates that fluid at the center of tube flows with a phase lag of 90° relative to the wave of the pressure gradient. In addition at this location the amplitude diminishes with increasing frequency. On the other hand expression (3.4.9) shows that the maximum velocity occurs in the neighbourhood of the wall for this case of a very high frequency. A few examples of the simple periodic pulsations given by

$$-\frac{1}{\rho}\frac{\partial p}{\partial z} = P_0 + P_{cn} \cos nt \tag{3.4.10}$$

are provided in Fig. 3.4.1 through Fig. 3.4.4.

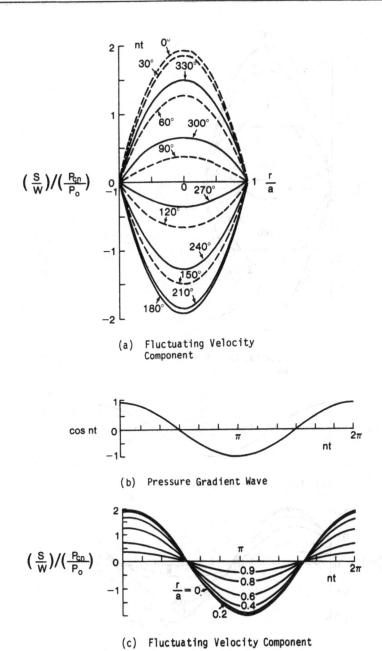

(a) Fluctuating Velocity
     Component

(b) Pressure Gradient Wave

(c) Fluctuating Velocity Component

Fig. 3.4.1 Velocity Profile for (ka) = 1. After S. Uchida, 1956.
            With permission.

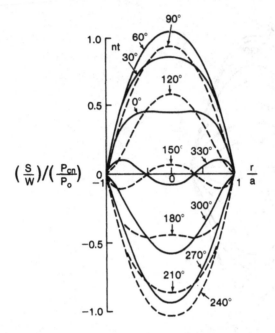

(a) Fluctuating Velocity
     Component

(b) Pressure Gradient Wave

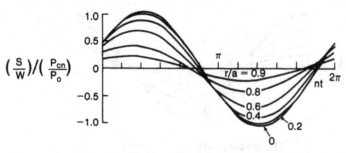

(c) Fluctuating Velocity Component

Fig. 3.4.2 Velocity Profile for (ka) = 3. After S. Uchida, 1956.
            With permission.

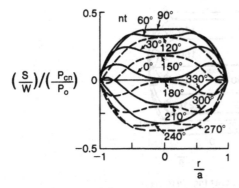

$\left(\dfrac{S}{W}\right)\Big/\left(\dfrac{P_{cn}}{P_o}\right)$

(a)   Fluctuating Velocity
        Component

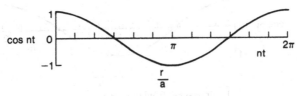

cos nt

(b)   Pressure Gradient Wave

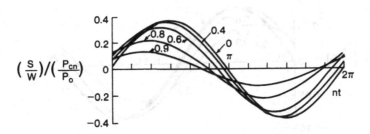

$\left(\dfrac{S}{W}\right)\Big/\left(\dfrac{P_{cn}}{P_o}\right)$

(c)   FLuctuating Velocity Component

Fig. 3.4.3   Velocity Profile for (ka) = 5.   After S. Uchida, 1956.
                  With permission.

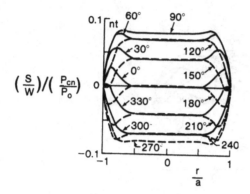

(a) Fluctuating Velocity
     Component

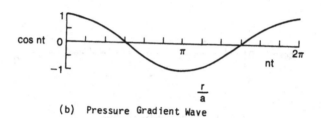

(b) Pressure Gradient Wave

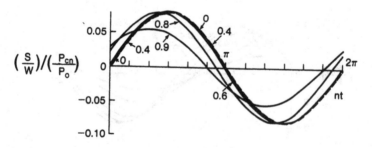

(c) Fluctuating Velocity Component

Fig. 3.4.4 Velocity Profile for (ka) = 10. After S. Uchida, 1956.
            With permission.

The solution obtained may be used in finding the instantaneous shear stress acting on the wall. This is given by $\mu \, \partial u / \partial r$, which is to be evaluated at $r = a$. Furthermore, the instantaneous drag may be found by integrating the shear stress over the area of interest.

3.5 Entrance Length

In the previous sections we have considered the fully developed flows. However, the arterial passages are of such geometrical config- uration that, although the fully developed flow considerations help in understanding the basics of the flow phenomena, the actual flow character- istics will be somewhat altered.

To illustrate: Consider a straight tube with a well-rounded entrance where the velocity profile is nearly uniform. Owing to viscous effects, a boundary layer will be formed on the wall. At some infinite distance downstream the boundary layer reaches the center of the pipe. Shortly thereafter the velocity at the center of the tube reaches 99 percent of the maximum of its final parabolic profile. This constitutes the en- trance length, ℓ. The flow from then on is considered to be fully de- veloped and is assumed to follow Poiseuille equations. According to White (1974), such condition is reached when

$$\ell/D \approx 0.08 \, \text{Re} + 0.7 \tag{3.5.1}$$

The volume of flow must be the same for every cross-section of the rigid tube. Consequently, within the entrance length, the decrease in the rate of flow near the wall which is due to viscous effects must be compensated by a corresponding increase near the axis. Thus the boundary layer in this case is formed under the influence of an accelerating external stream. Such a steady state flow will never separate from a regular wall and the shear stress at the wall will transform asymptotically from its initial value to the value which would exist if Poiseuille flow could be established. In the pulsatile case adverse pressure gradients may be present and, depending on fluctuation frequency and amplitude, reversed flow could appear. The associated minimum wall shear stress would be equal to zero and the wall drag would periodically change its direction.

In view of the above it may be recognized that in the large arteries, where blood may be assumed to behave like a Newtonian fluid, the fully developed flow conditions most likely does not exist.

3.6 Influence of the Blood Cells

If one considers that in blood there would be normally more than 5×10^3 million red cells per 1 mℓ occupying 40 to 45 percent of the total volume, it becomes necessary to investigate the influence of the presence of these particles on the flow characteristics. Though the other blood particles, at times, produce important effects in the flow characteristics, the red cells are so numerous that they largely determine the macroscopic flow pattern. There should be no doubt, that in order to reach a full understanding of the phenomena the effect of the presence of the corpuscles should be considered.

Many investigators disregard the action of the individual particles and consider blood as being a continuous substance and adopt a continuum model of fluid. Such a model possesses appropriate continuum properties

which are so defined as to ensure that, on the macrosopic level, the behavior of the model duplicates that of the real fluid. The mean properties of the fluid element are in the limit assigned to a point so that
we may ultimately adopt a field representation for the continuum properties. This is not necessarily always the case in arterial blood flow.
However, at least in the large vessels the assumption of a pure liquid
is a sufficiently close approximation. The continuum assumption is made
particularly inaccurate by the presence of a cell-depleted or cell-free
layer at the wall of blood vessel.

In essence there exists a mechanism which causes radial migration of
suspended corpuscles. This migration is predominantly in the direction
away from the wall. However, migration out from the tube axis has been
also observed. It has been attributed to the fact that a cell may possess
a degree of rigidity. In such circumstances, depending on the magnitude
of the Reynolds number, there appears to exist an equilibrium radial
position at which migration across lines of flow ceases (tubular "pinch
effect").

According to Goldsmith (1972) there are three main effects which
may be noted at high particle concentrations: (a) the velocity distribution is no longer parabolic (the profile becomes blunted in the tube
center where there is a region in which particles move with the same
speed); (b) the particle path has an erratic component in the direction
normal to the flow; (c) particle deformation in blood occurs to a degree
which cannot be attributed to shear alone.

There are indications that similar flow characteristics are present

in pulsatile flows as in steady flow. However, we may expect the magnitudes to change. For example, the "equilibrium position" may move
closer to the wall. Nevertheless, what actually occurs in vivo is still
left to speculation.

3.7 Arterial Pressure-Flow Relationship

The arterial pressure-flow relationship is the central problem in
haemodynamics. The driving force is the pressure gradient. Most of the
time fluid will flow in the direction of favorable pressure gradient
(decreasing pressure). However, at times it is possible for the fluid
to continue flowing against adverse pressure gradient (increasing pressure), until the kinetic energy of a particle in that direction is depleted. That is to say that the flow curve may be lagging behind the
pressure curve.

The timing of flow reversal was shown by D.A. McDonald (1955). Fig.
3.7.1 reproduces a volume flow curve and the pulse pressure in the femoral
artery of a dog. The femoral artery was chosen because more reports were
available on the flow pattern in this artery of the dog than on any other.
Pressure was measured by the use of capacitance manometers, recording
through a tube inserted into branches of the femoral artery and adjusted
so that the ends lay flush with the wall of the main vessel. Direct
measurements of flow have been made by following the movement of injected
bubbles of oxygen recorded by high-speed cinematography. The bubble fills
the tube and travels at the mean velocity.

It is shown in Fig. 3.7.1 that the maximum favorable pressure gradient is reached during the rising phase of the pulse wave at point a.

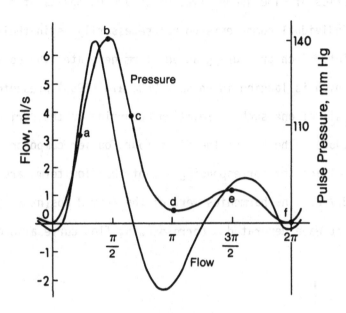

Fig. 3.7.1 The pulse pressure and the volume flow curves in the femoral
            artery of a dog. After McDonald, 1955. With permission.

However, due to fluid inertia and possibly wall effects the maximum forward rate of flow appears a little later. From point a to point b the
favorable pressure gradient decreases from its maximum value to zero,
which sets demand on the fluid to decelerate. Between points b and d the
adverse pressure gradient goes from zero to zero with its maximum value
at point c. This pressure gradient is present for a sufficiently long
time to reverse the flow. Here again, the maximum negative flow is delayed with respect to the maximum point of adverse pressure gradient.
Similar response of the flow curve to the pulse pressure curve is for the
d-e-f secondary diastolic pressure hill.

Curves such as those shown in Fig. 3.7.1 may be represented mathematically by a series of sine shape waves which are harmonics of a funda-

mental wave. Individual components do not necessarily begin their cycle

at the zero point. Each pressure gradient term generates the correspond-

ing flow term which is lagging in phase. McDonald (1955) presented a

graphical analysis of one such correlation according to the derivations

of Womersley (1955). The sum of the first four Fourier components of the

pressure gradient and the corresponding sum of the flow terms are repro-

duced in Fig. 3.7.2. The harmonics were of the form M cos(nt + ϕ). Each

pressure gradient wave generated a corresponding flow curve also of the

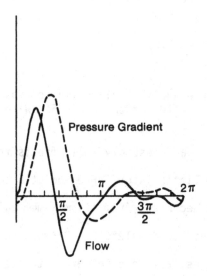

Fig. 3.7.2 The sum of the first four harmonics of the pressure gradient
            and the flow. After McDonald, 1955. With permission.

sine wave form but lagging in phase. Qualitative comparison of Fig. 3.7.1

and Fig. 3.7.2 indicates some disagreement in the latter part of the cycle.

However, inclusion of the fifth and sixth harmonics would move the two corresponding curves significantly closer.

In conclusion it was stated that in the femoral artery the flow oscillates in the .same way as the pressure gradient but with a phase lag which varies throughout the cycle. A large forward flow (7 to 15 times the mean flow rate) occurs during systole followed by a smaller back-flow phase and a subsequent forward flow during diastole.

4. FLOW IN CURVED PASSAGES

4.1 Introduction

Flow through curved arteries has been a subject of much concern in blood flow studies related to the atherosclerosis. Many parts of the human arterial tree are curved with various different radii. Martin (1960) reports the existence of S-shaped arteries in the vertebral cerebral system. Also it may be appropriate to refer here to the work of Mitchell and Schwartz (1965) who indicated that different segments of the same artery show considerable difference in disease severity. In parti-cular, in the "tortuous" arterial segments striking increases in severe disease have been found.

When a fluid flows through a curved passage a secondary flow occurs in planes perpendicular to the axis of the tube. For a given velocity profile some fluid particles possess a higher velocity than others. The centrifugal force created by motion around the bend, then, is larger for the faster moving particles than for the slower ones, leading to the emergence of a secondary flow directed outward in the centre and inward at the wall. Such flows were originally studied by Dean (1927), White (1929), Adler (1934), Nippert (1929), and Richter (1930). These experimentally show that the influence of curvature is stronger in laminar than in turbulent flow.

4.2 Fully Developed Flow in a Curved Pipe

At this time we will involve ourselves with the presentation of Cuming (1952). The right-handed orthogonal coordinates system has been

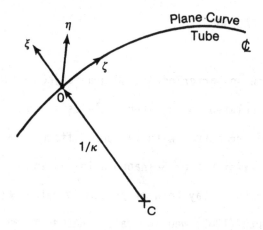

Fig. 4.2.1 Orthogonal System of Axes

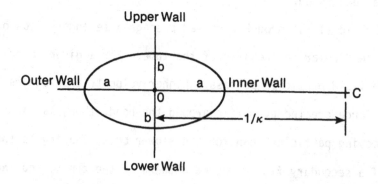

Fig. 4.2.2 Elliptic Section

used in connection with the assumed elliptical cross-section (2a and 2b major and minor axes, respectively): 0 ξ along the major axis; 0 η along the minor axis and 0 ζ along the curved central axis of the passage of the curvature κ (see Fig. 4.2.1 and 4.2.2.) The corresponding velocity components along these axes were taken as u, v and w, respectively.

Equations of motion could be presented in the dimensionless form by the following substitutions: ξ = ax, η = λay (λ = b/a), ζ = az, u^+ = au/ν, v^+ = av/ν, w^+ = aw/ν, p^+ = a^2p/(ρ ν^2). Substitution into Equations (2.2.1) through (2.2.4) yields

$$\frac{\partial}{\partial x}[(1 + \kappa ax) u^+] + \frac{1}{\lambda}\frac{\partial}{\partial y}[(1 + \kappa ax)v^+] = 0 \qquad (4.2.1)$$

$$u^+ \frac{\partial u^+}{\partial x} + \frac{v^+}{\lambda}\frac{\partial u^+}{\partial y} - \frac{\kappa a w^{+2}}{1 + \kappa ax} = -\frac{\partial p^+}{\partial x} + \frac{1}{\lambda^2}\frac{\partial^2 u^+}{\partial y^2} - \frac{1}{\lambda}\frac{\partial^2 v^+}{\partial x \partial y} \qquad (4.2.2)$$

$$u^+ \frac{\partial v^+}{\partial x} + \frac{v^+}{\lambda}\frac{\partial v^+}{\partial y} = -\frac{1}{\lambda}\frac{\partial p^+}{\partial y} + \frac{\partial^2 v^+}{\partial x^2} - \frac{1}{\lambda}\frac{\partial^2 u^+}{\partial x \partial y}$$

$$+ \frac{\kappa a}{1 + \kappa ax} (\frac{\partial v^+}{\partial x} - \frac{1}{\lambda}\frac{\partial u^+}{\partial y}) \qquad (4.2.3)$$

$$u^+ \frac{\partial w^+}{\partial x} + \frac{v^+}{\lambda}\frac{\partial w^+}{\partial y} + \frac{\kappa a u^+ w^+}{1 + \kappa ax} = -\frac{1}{1 + \kappa ax}\frac{\partial p^+}{\partial z} + \frac{\partial^2 w^+}{\partial x^2}$$

$$+ \frac{1}{\lambda^2}\frac{\partial^2 w^+}{\partial y^2} + \frac{\kappa a}{1 + \kappa ax}\frac{\partial w^+}{\partial x} - \frac{\kappa^2 a^2 w^+}{(1 + \kappa ax)^2} \qquad (4.2.4)$$

Approximate solution of Equations (4.2.1) through (4.2.4) have been obtained by Cuming (1952), by letting

$$u^+ = \kappa a Re^2 u'$$

$$v^+ = \kappa a Re^2 v'$$

$$w^+ = (\overline{w} + \kappa a Re^2 w' + \kappa a w'')\, Re/2$$

$$p^+ = \overline{p} + \kappa a Re^2 p'$$

$$Re = -\frac{\lambda^2}{1 + \lambda^2}\,\frac{\partial\, p^+}{\partial z}$$

where $\overline{p}$ and $\overline{w}$ are the pressure and axial velocity as indicated in Section 3.2 (flow through a straight elliptic pipe), respectively; and where Re is the Reynolds number for the flow also through a straight pipe. Equating coefficients of the zero and first power in κ, one obtains

$$-\frac{\overline{w}^2}{4} = -\frac{\partial p'}{\partial x} + \frac{1}{\lambda^2}\frac{\partial^2 u'}{\partial y^2} - \frac{1}{\lambda}\frac{\partial^2 v'}{\partial x \partial y} \tag{4.2.5}$$

$$0 = -\frac{1}{\lambda}\frac{\partial p'}{\partial y} + \frac{\partial^2 v'}{\partial x^2} - \frac{1}{\lambda}\frac{\partial^2 u'}{\partial x \partial y} \tag{4.2.6}$$

$$u'\frac{\partial \overline{w}}{\partial x} + \frac{v'}{\lambda}\frac{\partial \overline{w}}{\partial y} = \frac{\partial^2 w'}{\partial x^2} + \frac{1}{\lambda^2}\frac{\partial^2 w'}{\partial y^2} \tag{4.2.7}$$

$$0 = -2\left(1 + \frac{1}{\lambda^2}\right) x + \frac{\partial^2 w''}{\partial x^2} + \frac{1}{\lambda^2}\frac{\partial^2 w''}{\partial y^2} + \frac{\partial \overline{w}}{\partial x} \tag{4.2.8}$$

$$\frac{\partial u'}{\partial x} + \frac{1}{\lambda}\frac{\partial v'}{\partial y} = 0 \tag{4.2.9}$$

Eliminating pressure between Equations (4.2.5) and (4.2.6), and introducing stream function via Equation (4.2.9), velocity components u and v are obtained, namely

$$u = \frac{\kappa \upsilon Re^2}{\lambda^2} (1 - x^2 - y^2) [(1 - x^2 - y^2) (C_1 + C_2 x^2 + 3 C_3 y^2)$$

$$\text{(4.2.10)}$$

$$- 4y^2 (C_1 + C_2 x^2 + C_3 y^2)]$$

$$v = \frac{2\kappa \upsilon Re^2}{\lambda^2} (1 - x^2 - y^2) [2(C_1 + C_2 x^2 + C_3 y^2)$$

$$\text{(4.2.11)}$$

$$- C_2 (1 - x^2 - y^2)] xy$$

where

$$C_1 = \lambda^4 (375 + 820\lambda^2 + 1{,}114\lambda^4 + 212\lambda^6 + 39\lambda^8)/360(5 + 2\lambda^2 + \lambda^4) G(\lambda)$$

$$C_2 = -\lambda^4 (75 + 2\lambda^2 + 3\lambda^4)/360 \, G(\lambda)$$

$$C_3 = -\lambda^4 (15 + 26\lambda^2 + 39\lambda^4)/360 \, G(\lambda)$$

$$G(\lambda) = 35 + 84\lambda^2 + 114\lambda^4 + 20\lambda^6 + 3\lambda^8$$

Equations (4.2.10) and (4.2.11) indicate a set of streamlines as indicated in Fig. 4.2.3. The secondary flow consists of two opposed vortex

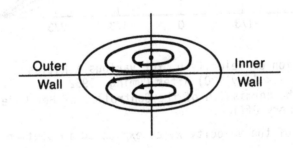

Fig. 4.2.3 Secondary Flow

motions in the top and bottom halves of the tube. In addition it has
been found that the pressure along the minor axis is equal to the arith-
metic mean of the pressures at x = 1 and x = -1. Furthermore, it has been
demonstrated that for large and small magnitudes of the λ ratio the secondary flow diminishes from its maximum value at λ = 2.2. Variation of
the velocity component u across the central plane (-1 $\leq$ x $\leq$ 1, y = 0) for
various λ ratios is shown in Fig. 4.2.4.

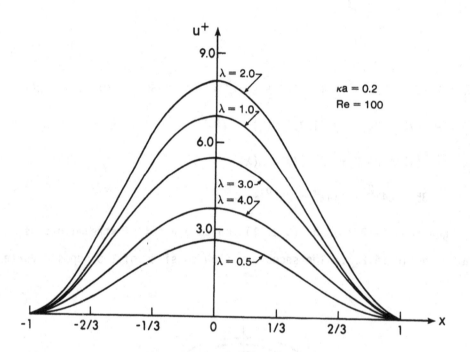

Fig. 4.2.4   Variation of Velocity u[+] for various λ Ratios
             (- 1 < x < 1, y = 0) After Cuming, 1952
             With the permission of the Controller of Her Majesty's
             Stationery Office.

        Substitution of the velocity $\bar{w}$, as expressed in section 3.2, into
Equations (4.2.7) and (4.2.8) yields solutions for w' and w''. These indicate that, to the degree of approximation considered, the axial velocity

component is modified by two curvature terms: the first of these w',
associated with the square of Re, causes the velocity to increase in the
outer half of the bend and to decrease in the inner half of the bend (effect associated with flow in a curved pipe); the second term w'' has reversed influence (effect associated with the flow in a curved channel).
For values of λ around unity the influence of the first term is predominant. However, with increasing λ the second term begins to dominate.
The associated results are shown in Fig. 4.2.5. It can be seen that as λ
increases the point of maximum axial velocity shifts from the outside of

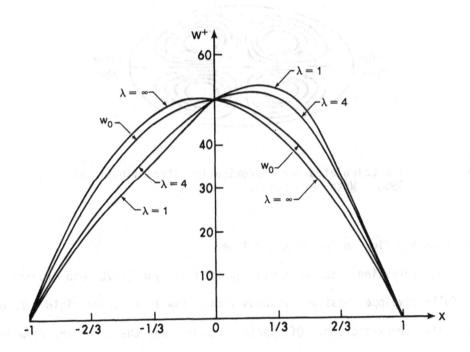

Fig. 4.2.5 Variation of Velocity w^+ for Various λ Ratios
            $(-1 < x < 1, y = 0)$ After Cuming, 1952
            With the permission of the Controller of Her Majesty's
            Stationery Office.

the bend over to the inside of the bend. For the special case when λ
ratio is unity we obtain the solution for the flow through a curved pipe
of circular section.

The above solution represents the first order modification to the
zero order approximation which is the same as the flow in a straight tube.
Ito (1950) obtained the second order modification which is presented in
Fig. 4.2.6. In practice the secondary flow of this order is added to
the secondary flow of the type shown in Fig. 4.2.3.

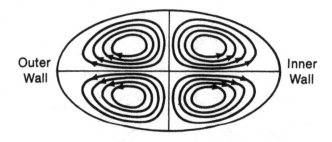

Fig. 4.2.6 The Calculated 2nd Approximation Streamlines. After Ito,
            1950. With permission.

4.3 Unsteady Flow in Curved Rigid Tube

An interesting phenomenon was reported by Lyne (1970) who studied
the fully developed unsteady viscous fluid flow in a curved rigid tube of
a circular cross-section. Of special interest was the secondary flow induced in the plane of the cross-section of the tube. In order to simplify
the problem, the radius of pipe curvature was assumed to be large in relation to its own radius, and the pressure gradient was assumed to be

sinusoidal in time with zero mean.

Lyne defined a nondimensional parameter β given by

$$\beta^2 = 2\nu/(na^2) \tag{4.3.1}$$

which may be interpreted as the ratio of Stokes' layer of thickness $(2\nu/n)^{1/2}$ to the radius of the pipe (note that $\beta = \sqrt{2}/\alpha$). It was found that for the sufficiently small values of β (this implies that the viscous effects are confined to a thin layer next to the wall, while the core of the flow is inviscid) the secondary flow in the central zone of the tube is in the opposite sense (see Fig. 4.3.1) to that predicted by the steady state analysis (see Fig. 4.2.3). That is the secondary flow due to the centrifugal effect is confined to the thin boundary layer region, which in turn drags the fluid in the central region.

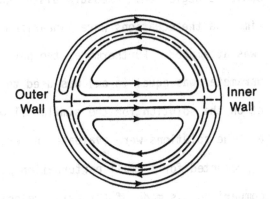

Fig. 4.3.1 The Streamlines for Small β. After Lyne, 1970. With permission.

4.4 Unsteady Flow in Thin-Walled Curved Elastic Tube

In the study of etiology of arterial diseases it is relevant to understand the mechanics of pulsatile flow not only in rigid tubes but also in elastic tubes. Chandran et al (1974) presented their investigation on the oscillatory flow in thin-walled tubes which correlates with findings of Lyne (1970). This work is the object of the present section.

Equation (2.3.1) through (2.3.4) have been written in the toroidal coordinate system and supplemented by the corresponding equations of motion for the tube. To include the effect of the surrounding tissues on the arteries, a spring-mass restraint has been introduced. Due to the assumption of small radial tube displacements the boundary conditions were linearized by equating the velocity components at the undisturbed tube radius rather than at the instantaneous tube radius.

An order of magnitude study has shown that a number of terms in the governing equations could be neglected. Pressure disturbance was assumed to be sinusoidal in time and the equations were linearized before a wave propagation analysis was attempted. Furthermore, the perturbation technique (perturbation parameter was equal to aκ) was used to introduce the small curvature effect on the solution for the similar flow through a straight elastic tube. The solutions were evaluated numerically.

The results were presented for a small perturbation parameter, namely aκ = 0.1. A comparison was made of the axial velocity profile of the oscillatory flow in a straight elastic tube with that of curved elastic tube. In conclusion it was reported that the maximum axial velocity, for the case of time dependent flow in curved tube may be shifted towards the center of curvature. This shift is opposite to that for the steady

flow in curved tubes (see Fig. 4.2.5).

4.5 Entrance Arc.

The fully developed flow in a curved passage is defined as the

region in which the velocity no longer depends upon the distance in the

axial direction. It is the outcome of the flow reconfiguration which

takes place within the entrance arc. In a curved tube, and for a steady

state, this entrance arc is measured by the number of degrees around the

curve from the tube entrance to the location where the fully developed

flow has established itself. Scarton et al (1977) quote the following

formula for the entrance arc:

$$49 \ (Re)^{1/3} \ (a \ \kappa)^{1/2} \qquad\qquad\qquad (4.5.1)$$

It may be noted that, for example, in the human aortic arch the en-

trance arc, most of the time, will be in excess of the angular distance

from the aortic root to the branching site at the upper surface of the

aortic arch. It will be also recognized that in our arterial system

there are present numerous successive branching sites. In addition the

axis of curvature of practically any arterial tube is not necessarily in

the same plane. Consequently, it is most unlikely that in the arterial

tree of the human the fully developed flow could exist.

4.6 Laminar Flow Downstream of a Bend

Measured and calculated velocity profiles were reported for the

water flow downstream of a 15 degree bend in a round glass tube at a

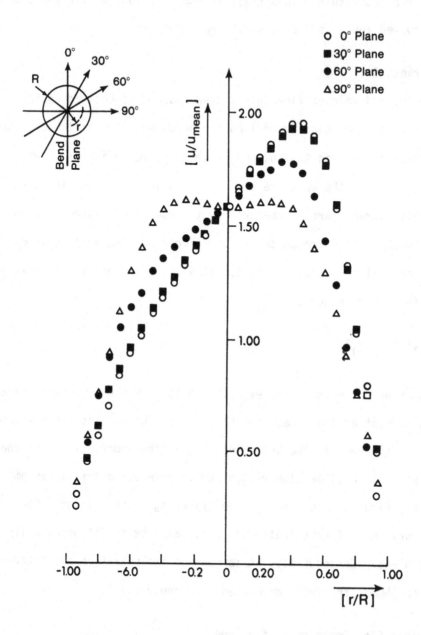

Fig. 4.6.1   Experimental Velocity Profile at Initial Section
After Gosman et al., 1975. With permission.

Reynolds number of 300 by Gosman et al (1975). As water is a Newtonian

fluid and blood flowing in the large arteries behaves like a Newtonian

fluid, this investigation is pertinent to the present subject.

The measurements were obtained using laser-Doppler anemometry and

provided information regarding the development of the axial velocity from

the exit of the bend to the fully developed situation. Initial flow at

the entrance to the bend was fully developed. The results are shown in

Fig. 4.6.1 through 4.6.3. The calculation procedure was based on a finite-

difference solution of the three-dimensional, steady, boundary-layer form

of the Navier-Stokes equations given by Eqn. 2.2.1 through 2.2.4. Com-

parison between the measurements and calculations revealed close agreement

and indicated the high precision of both calculation and measurement tech-

niques. The maximum deviation of the calculated velocity values from the

measured quantities was 2.5%.

Fig. 4.6.1 shows the measured velocity profiles across four diameters

at the initial station which was located approximately 5 tube diameters

downstream of the centre of the bend. These values were used as initial

conditions for the solution of the appropriate governing equations. Note

the double peak velocity profile for the plane perpendicular to the plane

of the bend. Fig. 4.6.2 shows velocity values computed and measured across

diameters normal to the plane of the bend and at six consecutive axial

locations. These profiles are symmetric and indicate that the fully-

developed condition was reached within 20 tube diameters of the centre of

the bend or within 15 tube diameters of the initial station. The velocity

profiles computed and measured in the plane of the bend are shown in

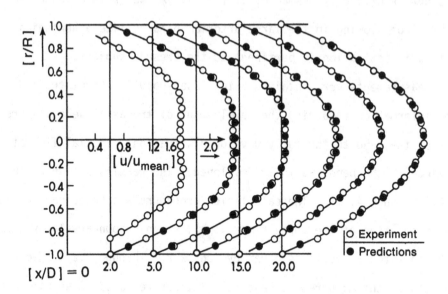

Fig. 4.6.2   Flow Development in Normal to the Bend Plane
             After Gosman et al., 1975. With permission.

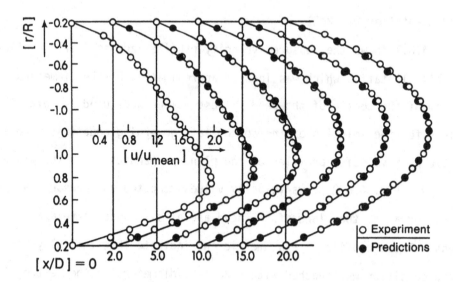

Fig. 4.6.3   Flow Development in the Bend Plane. After Gosman et al.,
             1975. With permission.

Fig. 4.6.3 As could be expected the maximum of the velocity is shifted towards the outer wall of the bend. Again here the fully-developed flow is reached within 20 tube diameters of the centre of the bend.

4.7 Drag on the Wall

A vector addition of velocity components yields the magnitude and direction of the absolute velocity of a participating fluid particle. These give the velocity profiles and in turn the shear stresses which can be computed with the aid of the expressions provided in Section 2.4. Integration of the shear stresses, over some specified area, produces the local time dependent drag which at the wall assumes the following form

$$\text{Drag} = \int \tau_0 dA \qquad\qquad\qquad\qquad (4.7.1)$$

where τ_0 is the shear stress evaluated at the wall and A is the wall area. To overcome such a flow resisting drag a driving force in the form of a corresponding pressure gradient must be applied.

5. FLOW IN JUNCTIONS

5.1 Simple Bifurcation

The manner in which the flow divides at a simple arterial junction (where a single branch leaves the straight-through parent tank) was studied in terms of the significant dimensionless parameters by Rodkiewicz and Howell (1971). They found that, for certain values of these parameters, more fluid goes into the side branch than straight through. In order to have a better understanding of this phenomenon and in order to find why and when the mass flow ratio γ (the ratio of the rate of flow in

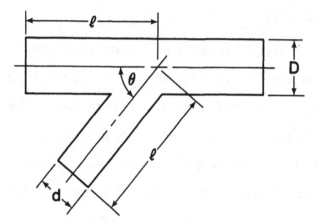

(a)   With Sharp Inside Edges

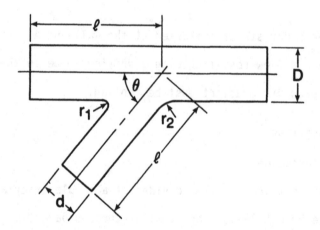

(b)   With Rounded Inside Edges

Fig. 5.1.1 Experimental Bifurcations

he side branch to the flow rate in the main branch) becomes greater than

ne, the study of the steady-state flow characteristics in an arterial

unction has been undertaken by Rodkiewicz and Roussel (1973). Their

ange of study was: $1000 \leq Re \leq 5000$ and $0.40 \leq \beta \leq 1$ (β = d/D; see Fig.

.1.1). In order to enable observations and photography, the bifurcations

ere made of acrylic. Sixteen junctions were tested; $\bar{\theta} = \theta/90 = 1/3$,

/9, 7/9, 1, and β = 0.4, 0.6, 0.8, 1.0. The isolated basic geometry

epresents the usual idealization of the actual arterial flow problem.

lthough the bifurcations inside the body generally have a tapered main

essel and a possibly more gradual curvature to the side branch, these

dditional variables were not considered. Hydrogen bubble technique was

sed to visualize the flow distribution within the bifurcation. The

lectrolysis direct current was brought to the electrodes from the gener-

tion unit which allowed the operator to have control of the frequency and

he duration of the pulses producing the bubbles.

The dependence of the mass flow ratio γ on the pertinent dimension-

ess parameters is reproduced in Fig. 5.1.2 through 5.1.5. These graphs

ndicate that the γ ratio decreases when the Reynolds number increases.

t could be attributed to the inability of the fluid to negotiate the

urn, due to an increase in its momentum in the mainline direction. The

raphs also give the variations of γ with diameter ratio β and the angle

atio $\bar{\theta}$.

It has been established that, for the equal resistance discharge, a

hange in the γ ratio is due to a variation in size and location of two

ndependent (one in the main branch and one in the side branch) separation

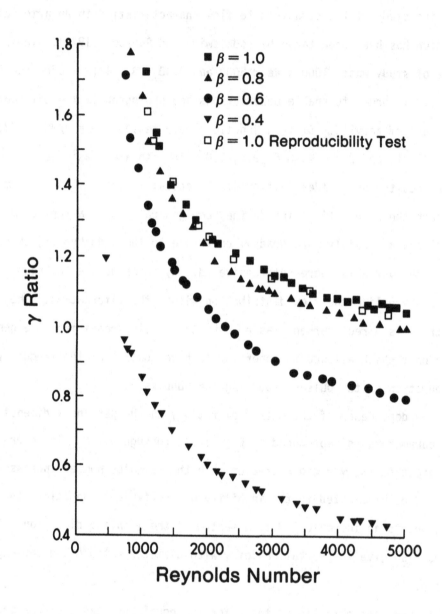

Fig. 5.1.2 Mass Flow Ratio γ Versus Reynolds Number Re as a Function
            of β for θ̄ = 1/2 (θ = 30 deg)

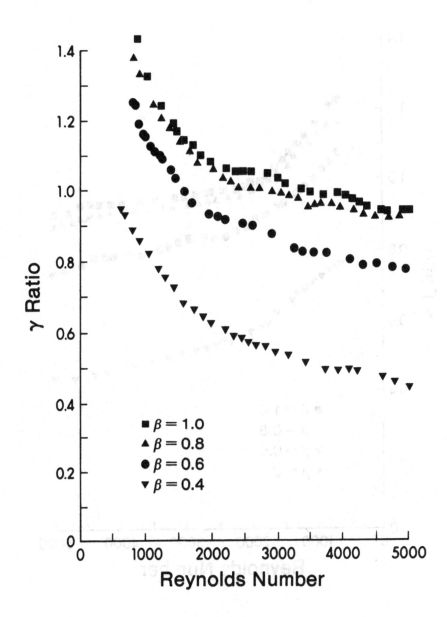

Fig. 5.1.3 Mass Flow Ratio γ Versus Reynolds Number Re as a function
of β for $\bar{\theta}$ = 5/9 (θ = 50 deg)

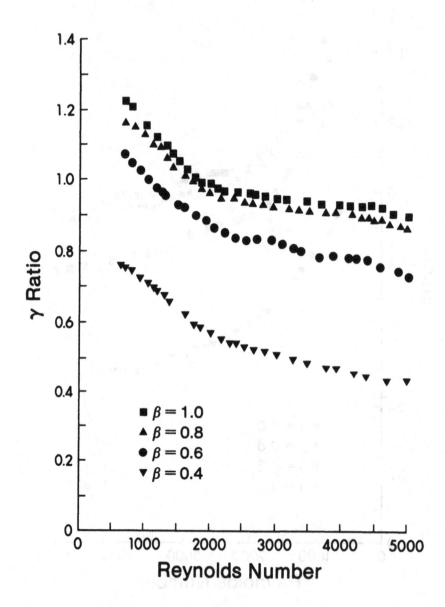

Fig. 5.1.4   Mass Flow Ratio γ Versus Reynolds Number Re as a Function
           of β for θ̄ = 7/9 (θ = 70 deg)

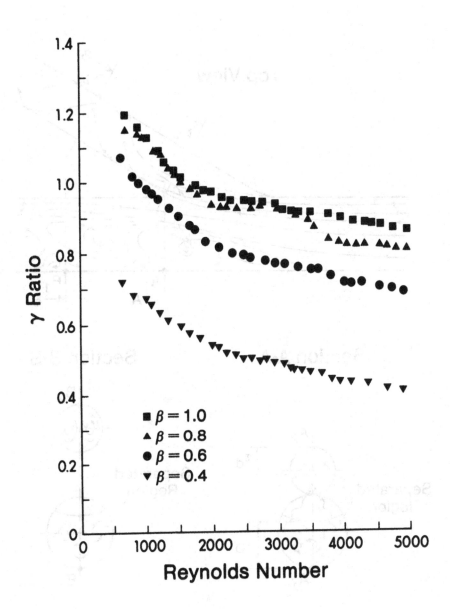

Fig. 5.1.5 Mass Flow Ratio γ Versus Reynolds Number Re as a Function
of β for $\bar{\theta}$ (θ = 90 deg).

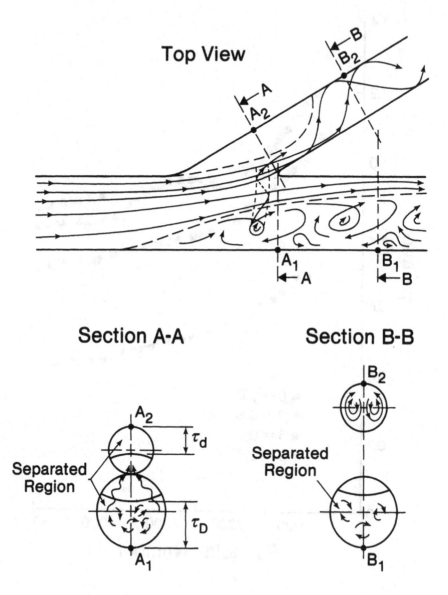

Fig. 5.1.6 Physical description of the flow. After Rodkiewicz and
            Roussel (1973).

regions. This is shown in Fig. 5.1.6. Defining the thicknesses of the separation regions by τ_d and τ_D (Fig. 5.1.6), one can observe that, whenever τ_D increases, τ_d decreases, and vice versa (see Fig. 5.1.7 through 5.1.9). Since growth of the separation region in the main branch restricts the flow in this branch, the mass flow ratio increases. The separation region in the main branch was found to grow when the Reynolds number decreases (Fig. 5.1.7), when the diameter ratio increases (Fig. 5.1.8) and when the angle of branching decreases (Fig. 5.1.9). At lower Reynolds numbers this separation region is so thick that the γ ratio becomes greater than 1, i.e. more fluid goes into the side branch than via the straight-through main branch.

Inside the separation region of the main branch vortices are generated. This is shown in 5.1.6 and 5.1.10. The vortices have been observed to be "banana-shaped". This shape enables, in the early life of a vortex, some particles to be transferred from the main-branch separation region into the side branch, as sketched in Fig. 5.1.6. Inside the side branch the flow becomes double-helicoidal, which is typical of flows in curved pipes.

The transition from laminar flow to turbulent flow may occur for a Reynolds number greater than 2000. The turbulence, flattening the velocity profile in the main branch, enables more fluid to flow into the side branch than in the laminar case for the same Reynolds number. This increases the mass flow ratio γ. The effect is particularly pronounced for large values of $\bar{\theta}$ and β, as can be seen in Fig. 5.1.5.

It may be of interest to mention that Roussel et al. (1973)

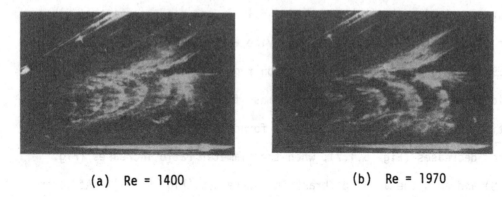

(a) Re = 1400   (b) Re = 1970

Fig. 5.1.7 Reynolds number dependence (β = 1.00, $\overline{\theta}$ = 1/3, θ = 30 deg)

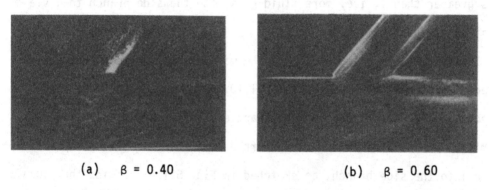

(a) β = 0.40   (b) β = 0.60

Fig. 5.1.8 Diameter ratio dependence ($\overline{\theta}$ = 5/9, θ = 50 deg, Re = 1600)

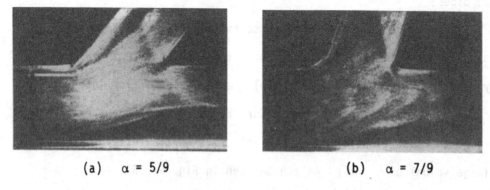

(a) α = 5/9   (b) α = 7/9

Fig. 5.1.9 Angle of branching dependence β = 0.80, Re = 1160

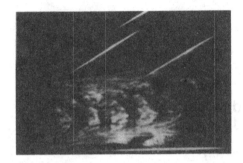 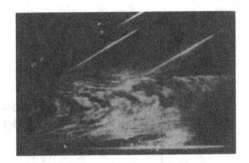

Fig. 5.1.10 Vortex formation ($\overline{\theta}$ = 1/3, θ = 30 deg, β = 1.00, Re = 1520)

investigated the question of the importance of the edge roundness at the

junction (see Fig. 5.1.1). His range of study was 1000 $\leq$ Re $\leq$ 5000, and

it was found that earlier conclusions (Rodkiewicz and Roussel, 1973)

were valid whether the bifurcation had sharp or rounded edges. However,

for Reynolds numbers less than or equal to 2000 (see Fig. 5.1.11 and

5.1.12) γ is slightly higher when the inside edges of the bifurcation

are rounded. This is due to the fact that the point of separation in

the side branch is pushed slightly downstream in the case of the rounded

edge bifurcation.

It has been emphasized (Rodkiewicz and Hsieh, 1975) that flow

characteristics for free discharge cases have significance predominantly

as an experimental base state. A full understanding of such a state is

helpful in analyzing the effect of various kinds of added resistances

downstream of the junction. Such is the case in blood flow where the

junction flow properties are usually controlled by the system which

follows the dividing element. Rodkiewicz and Hsieh (1975) demonstrated

one of these cases by extending the straight-through section on one of

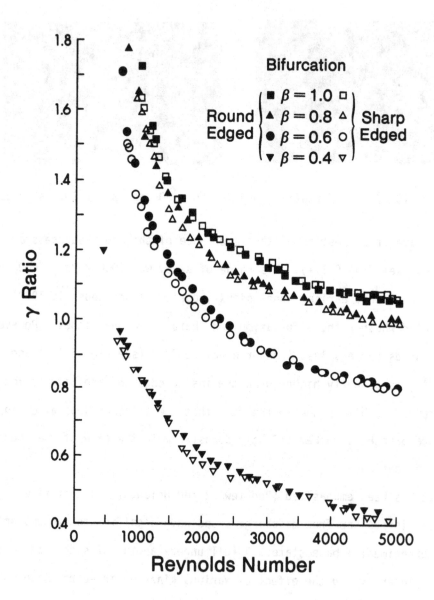

Fig. 5.1.11 Mass Flow Ratio γ Versus Reynolds Number Re as a Function
of β for θ̄ = 1/3 (θ = 30 deg)

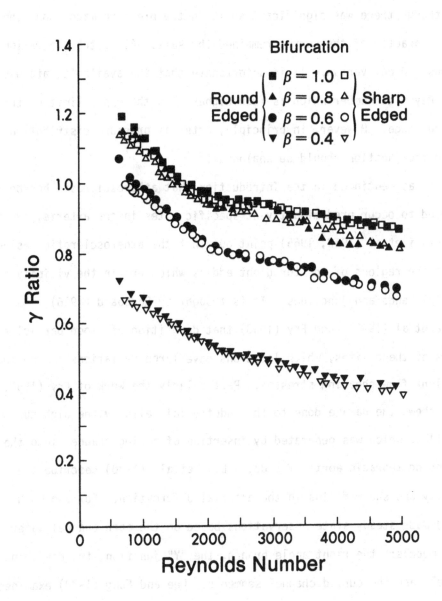

Fig. 5.1.12 Mass Fow Ratio γ Versus Reynolds Number Re as a Function
of β for $\bar{\theta} = 1$ (θ = 90 deg)

their junctions, i.e. adding resistance to the main flow. It was found

that though there was significant shift in the pressure magnitude, the

basic character of the curves remained the same. Also, the separation

regions did not vanish. It is unfortunate that the available data was

for a Reynolds number which is much higher than the upper limit of the

desired range. However, in principle, arterial pressure distribution

within the junction should be analogous.

It was mentioned in the Introduction that atherosclerosis has been

observed to occur predominantly at specific sites in the arterial system.

Wesolowski et al (1962, 1965) point out that the atherosclerotic lesions

are in the regions of the turbulent eddies which are in the vicinity of

arterial bends and junctions. It is thought by Rodboard (1956),

Downie et al (1963), and Fry (1968) that deposition of blood particles

occurs at these sites, which locations have large variations in the local

turbulent flow shearing stresses. Particularly the work of Fry (1968,

1969) shows the damage done to the endothelial cells by the high turbu-

lent flow, which was generated by insertion of a plug channel into the

descending thoracic aorta of a dog. Lynn et al. (1970) describe the

velocity and shear fields in the arterial bifurcation. Fox and Hugh

(1966) made steady state observations based on the open channel water

table models: the right angle branch, the "Y" junction, the divergent

channel, and the curved channel segments. Lee and Fung (1970) examined

the steady laminar flow through a tube with an axisymmetric constriction.

Just downstream of the constriction they reported a region of reverse

flow. This region was thought to be a significant factor in the deposi-

tion process which has been studied by Friedlander and Johnstone (1957).
One of the good examples associating fluid flow distribution with the
atherosclerotic plaques distal to orifices of the intercostal arteries in
dogs was presented by Texon (1972).

5.2 Symmetric Y Junction

There have been numerous studies dealing with the symmetric Y junctions. One of these was done by Kandarpa and Davids (1976). Though the
focus of their work was on the two-dimensional flow, the flow field they
presented (velocity, pressure and shear stress distribution) represents
qualitatively the situation which one may expect in the circular cross-
section bifurcations. In fact, it has been observed by Crow (1969) that
in principle the flow patterns are similar in circular and rectangular
geometries. However, it will be recognized that in the two-dimensional
passages one is not able to observe the dramatic doublehelicoidal patterns
and the double peak velocity profiles which may be present in a plane perpendicular to the plane of the Y junction (Brech and Bellhouse, 1973).

The typical steady state velocity profiles reported by Kandarpa and
Davids (1976) are shown in Fig. 5.2.1. A short distance into the tapered
segments, along the outer wall, the boundary layer separates and then reattaches. Variation of wall shear stress with the Reynolds number and
with the distance is shown in Fig. 5.2.2. It will be noted that there is
a change of sign of the outer wall shear stress as the flow goes through
the separation point and then when it reattaches. Along the branch inner
wall, a peak shear stress occurs at the leading edge (the smaller the bifurcation angle the greater the shear stress) and decreases to a value

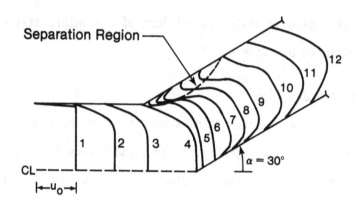

Fig. 5.2.1   Separation region in bifurcated channel.   After Kandarpe &
Davids, 1976.   With the permission of Pergamon Press Ltd.

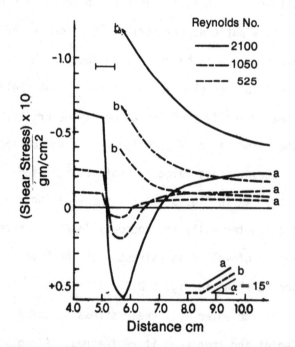

Fig. 5.2.2   Variation of wall shear stress with Reynolds number along
outer (a) and inner (b) walls. After Kandarpa & Davids, 1976.
With the permission of Pergamon Press Ltd.

which is attained along both walls. Modelling was made so that branch
dimensions were approximating those of the iliac of a dog. Prediction of
the leading edge shear stress was around 55 dyne/cm^2 for an entrance Reynolds number of 1050.

5.3 Note on Non-Rigid Wall

If we now relax the assumption of rigidity of the wall the interior
arterial surface will be subjected to pulls back and forth at the locations of the shearing stresses. This action, say around the indicated
separation arcs (see Fig. 1.3.1), will be repeated a great many times
(converging shear stresses). Restoring forces will be effective as long
as the tissue is resilient. However, if for some reason the tissues show
fatigue, creep could begin. One could draw a parallel here of a sagging
face. The effect of it would be creation of a ridge which is dramatically
illustrated in Fig. 5.3.1. This immediately extends the separation region

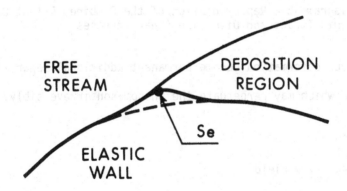

Fig. 5.3.1 Diagrammatic Representation of the Effect Due to Converging
Shear Stresses

and initiates the depository tendencies, should such tendencies not be
present before.

The diverging time and position dependent shear stresses at the
stagnation arcs (see Fig. 1.3.2) may bring the same effect, but on each
side of the arc. The tissue at the stagnation arc will be in tension at
all times and under the momentum flux of the impinging fluid molecules.
Possible wall deformation is indicated in Fig. 5.3.2. It may be noted

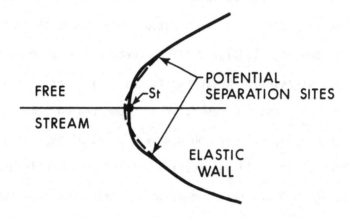

Fig. 5.3.2 Diagrammatic Representation of the Combined Effect Due to
            Fluid Impact and Diverging Shear Stresses

that should such deformation become permanent additional separation sites
may be created which may perpetuate the phenomenon irreversibly.

6. AORTIC ARCH

6.1 Aortic Arch Flow Field

Meisner and Rushmer (1962) reported steady state flow patterns observed in a flat rectangular model of the outflow tract of the left ventricle with rigid "open valves". The flow was essentially two-dimensional.
Existence of some separation regions was indicated. The knowledge of the

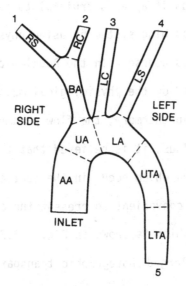

Fig. 6.1.1 Proposed Shape of the Aortic Arch Model

separation and stagnation regions within the aortic arch, the site which

according to Spain (1966) is among those positions which are favored by

the atherosclerotic plaque, appeared to be important. Rodkiewicz (1975)

determined where such regions can be expected in the system with specific

conditions and configuration of the aortic arch of the human, and fol-

lowed it with the experimental localization of the early atherosclerotic

lesions in the rabbit. The shape of the open channel model of the aortic

arch used (AA = Ascending aorta, BA = Brachiocephalic artery, LA = Lower

aortic arch, LC = Left common carotid, LS = Left subclavian, LTA = Lower

thoracic aorta, RC = Right common carotid, RS = Right subclavian, UA =

Upper aortic arch, UTA = Upper thoracic aorta) is shown in Fig. 6.1.1.

Research into the detailed flow patterns in the human aortic arch is

difficult to perform in vivo. Similarly, due to the complexity of the geometry, the flow through the region does not lend itself easily to theoretical studies. Thus it appears fruitful to gain the basic knowledge of the flow in such a branching system by using physical models. Several studies have followed this course. In the Rodkiewicz (1975) study, the model has a geometry based on the physiological measurements, even though it is two-dimensional. Furthermore, the flow parameters could be con-trolled independently. Thus, it is believed that this model simulated some of the flow phenomena which occur in the aortic arch region.

At this point it is convenient to present the average shape of the aortic arch of a rabbit which is shown in Fig. 6.1.2. It has been deter-mined by superimposing enlarged photographic transparencies of a number of rabbits studied in the project (Rodkiewicz, 1975). A comparison of Fig. 6.1.1 and Fig. 6.1.2 indicates a great similarity in the geometry of the

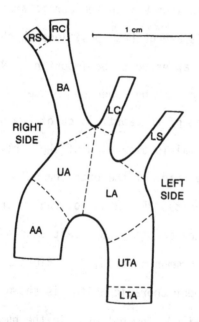

Fig. 6.1.2. Aortic Arch of the Rabbit.

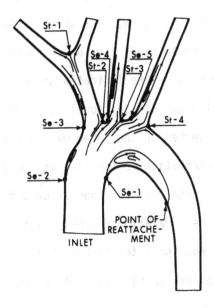

Fig. 6.1.3. Experimental Flow Pattern in the Aortic Arch Model

aortic arches of humans and rabbits. Consequently, flow similarities
should also exist. In addition, under certain laboratory conditions
rabbit lesions imitate relatively quickly the advanced stages of human
atherosclerosis (Constantinides, 1965).

Experimental results related to the flow pattern are diagrammatically
shown in Fig. 6.1.3. At each of the upper aortic branch (BA, LC, LS)
entrance sites there is one stagnation point (St-2, St-3, St-4), and one
separation point (Se-3, Se-4, Se-5) followed by a separation region where
the fluid turns back in the layer next to the wall. Separation and stagnation points are on the right and left walls, respectively. These points
actually represent segments of the circumferential length in the 3-dimen-
sional model. The phenomenon was isolated as shown in Fig. 6.1.4. The

dark areas indicate fluid that is free of the suspended particles which here are in the process of evacuation. The white areas represent the separation regions. One should study Fig. 6.1.4 in conjunction with Fig. 6.1.3 where all the regions are indicated graphically.

In the 2-dimensional model an additional very shallow separation region, Se-2, has been observed on the outer wall and at the end of AA. Furthermore, there is a stagnation region St-1 at the junction of RS and RC. It is interesting to note that the generation and shedding of vortices is present at all the separation regions. These are vividly depicted in Fig. 6.1.4 but not indicated in Fig. 6.1.3.

Opposite to the upper branching sites, at the maximum curvature of

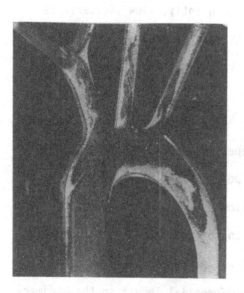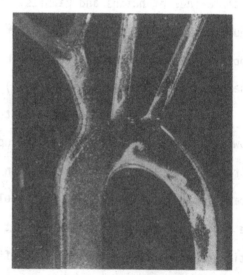

Fig. 6.1.4 Selected Flow Photographs

this two-dimensional model, there appears to be a relatively large separation region. It begins at the separation arc Se-1 at the end of AA,
and is terminated at the arc of reattachment within UTA. The reattachment site can be seen readily in the photographs of Fig. 6.1.4. In steady
flow, for a given Reynolds number, reattachment arcs oscillate to and fro
as the vortices are swept downstream. The vortex formation of this separation region is particularly well seen in Fig. 6.1.4. These reattachment arcs are rather difficult to locate, particularly in the side
branches where the separation regions are shallow. The flow and the shear
stress configuration in the reattachment sites are similar to those shown
in Fig. 1.3.2.

     With pulsatile flow the configuration, in principle, is the same.
Comparison of steady and pulsatile flow in a 2-dimensional model indicates
that the separation and stagnation regions exist in similar locations
most of the time. However, the location of the separation arcs now oscillate and the flow distribution within the region itself undergoes cyclic modifications. It was interesting to note that when the instantaneous
net flow rate was decreasing, the layers next to the wall, within the
separation regions, have been seen to move very swiftly in the upstream
direction.

     At this time it should be indicated that there has been a controversy
regarding existence of the separation region at the maximum curvature of
the aortic arch (i.e. separation region commencing at Se-1). The presently described observations were made on an open channel model (2-dim-
ensional). However, in a curved convergent tube (3-dimensional), like

the main trunk of the aortic arch, no separation should be present when there are no branches. On the other hand, sufficiently significant branching should generate this separation.

This hypothesis has recently been supported by steady and pulsatile flow tests in a 3-D model (Pelot and Rodkiewicz, 1982). For steady flow, there is evidence of separation at the point of maximum curvature when the Reynolds number exceeds 1000. The separation is dependent on the presence of the branches which act as diffusers, lowering the velocity in the arch and causing the adverse pressure gradient required for separation. For pulsatile flow, the backflow in the separation zone persists through-out the period. The 3-D model experiments also confirmed the existence of Se-2, Se-3, St-2, St-3 and St-4 (the other points not being considered).

6.2 Early Atherosclerotic Lesions

It has been postulated (Rodkiewicz, 1975) that the atherosclerotic formations within the aortic arch of the human may commence at the sites of possible separation and stagnation regions. As a consequence of the geometry, flow, and functional similarity, male rabbits were used in the experiment.

Animals were kept in separate cages and were allowed water and food as desired. The basic control diet was "Master Baby Rabbit" food (Maple Leaf Mills Ltd., Canada) containing, besides minerals and vitamins, 18% of crude protein and 2.5% of crude fat. To make the "cholesterol test diet" cholesterol was added to this basic diet, 5 g per 100 g of dry diet. Control rabbits were fed basic diet and sacrificed for the study of the aorta after the sampling of blood at the end of four weeks. "Cholesterol

fed" rabbits were on a cholesterol rich diet for periods of 3, 4, 5, 6, 7
and 8 weeks. At the end of a given feeding period, each rabbit had its
blood sampled from the median auricular artery and then was sacrificed by
the intra-arterial injection of sodium pentobarbital. The aorta of each
animal was then opened and prepared for photography. Serum cholesterol
was determined in the blood of control and cholesterol-fed animals following the method of Abell et al. (1952).

Average daily food and water consumption was respectively, 100 g and
194 ml per animal in control rabbits and 117 g and 195 ml in cholesterol
fed rabbits. In the first four weeks of experiments the body weight of
controls increased by 22.3% and of cholesterol fed animals by 22.8%. In
respect to food and water consumption and body-weight-increase two groups
of rabbits did not differ significantly. The cholesterol rich diet resulted in a significant increase of total serum cholesterol. The average
serum cholesterol was 38 mg/100 ml of serum in control animals. It increased to 1380 mg/100 ml of serum after three weeks of feeding on a
cholesterol rich diet. It averaged 2300 mg/100 ml of serum in animals
fed the cholesterol rich diet for 8 weeks.

The experimental results are reproduced in Fig. 6.2.1 through 6.2.6.
After three weeks of cholesterol rich diet the aortic arches indicated
some shadows in the surface, in comparison with the perfectly clear aortas
of the control rabbits. These shadows, as indicated in Fig. 6.2.1, have
been found exactly in all four stagnation and five separation regions
formed on the open channel model.

At the end of the fourth and fifth week, Fig. 6.2.2 and 6.2.3, it
was found that shadows extended to cover almost the upper half of the

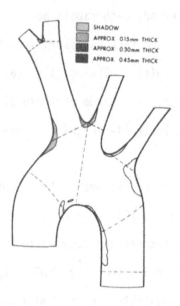

Fig. 6.2.1. Atherosclerotic Formations after 3 weeks on the diet.

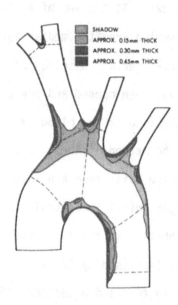

Fig. 6.2.2. Atherosclerotic Formations after 4 weeks on the diet.

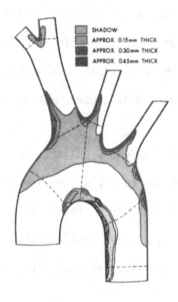

Fig. 6.2.3. Atherosclerotic Formations after 5 weeks on the diet.

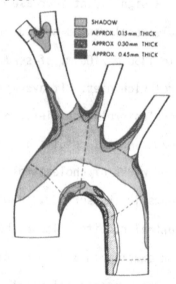

Fig. 6.2.4. Atherosclerotic formations after 6 weeks on the diet.

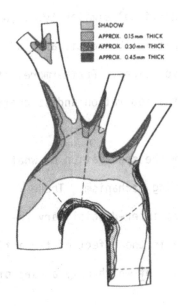

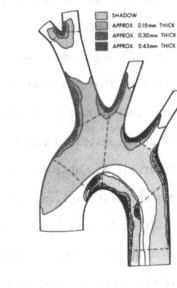

Fig. 6,2.5. Atherosclerotic formations after 7 weeks on the diet.

Fig. 6.2.6. Atherosclerotic
formations after 8 weeks on the
diet.

aortic arch and the rather long and narrow area at the aorta's maximum

curvature. However, at the locations where the shadows originated thicker

layers appear and grow as the time progresses. This can be seen in Fig.

6.2.4, 6.2.5, and 6.2.6 which represent rabbits after 6, 7 and 8 weeks on

the diet, respectively.

One notes that the atherosclerotic formations did not appear at the

sites where separation or stagnation regions were not expected. Further-

more, surprisingly the traces very closely reflect the flow distribution

reported in Fig. 6.1.4.

In summary the experimental evidence verified that atherosclerotic

formations in the aortic arch of the rabbit commence and develop at the

locations where there are expected separation and stagnation regions.
As a consequence of the geometry, flow and functional similarity, it was
assumed that the atherosclerotic formations within the aortic arch of
the human will begin and develop at similar locations. Furthermore, it
was speculated that the same rules apply to the separation and stagnation
regions of the various Y junctions.

The above described report refrains from the speculation of what is
atherosclerosis and what is its exact generating mechanism. These
questions merit special and substantially more inter-disciplinary research. It could well be that the phenomenon is the effect of the action-
reaction inter-relation of a number of elements of which only a part of
the full story was isolated and reported.

LITERATURE AND REFERENCES

Abell, L.L., Levy, B.B., Brodie, B.B. and Kendall, F.E. (1952), A simplified method for the estimation of total cholestral in serum and demonstration of its specificity, J Biol Chem, 195, 357.

Agrawal, Y.C. (1975), Laser velocimeter study of entrance flows in Curved
pipes, U. of California Berkeley Report FM-75-1.

Amyot, J.S., Francis, G.P., Kiser, K.M. and Falsetti, H.L. (1970),
Measurement of sequential velocity development in the aorta, ASME
Paper 70-WA/BHF-13.

Anderson, B. and Porje, I.G. (1946), Study of Ph. Broemser's manometer
theory for oscillations in the aorta, Acta Physiol Scand, 12, 3.

Atabek, H.B. (1964), End effects in Pulsatile Blood Flow, Attinger, E.O.,
ed., McGraw Hill, New York, 201.

Attinger, E.O. (1963), Pressure transmission in pulmonary arteries related to frequency and geometry, Circ Res, 12, 623.

Attinger, E.O. (1966), Hydrodynamics of blood flow, in Advances in

Hydroscience, vol. 3, Chow, V.T., ed., Academic Press, New York, 111.

Attinger, E.O., Anné, A. and McDonald, D.A. (1966), Use of Fourier series for the analysis of biological systems, *Biophys J*, 6, 291.

Austin, L.R. and Seader, J.D. (1973), Fully developed viscous flow in coiled circular pipes, *AICHE J*, 19, 85.

Back, L.D. (1975), Theoretical investigation of mass transport to arterial walls in various blood flow regions - I and II, *Mathematical Biosciences*, 27, 231.

Back, L.D., Radbill, J.R. and Crawford, D.W. (1977), Analysis of pulsatile, viscous blood flow through diseased coronary arteries of man, *J Biomech*, 10, 339.

Bard, P. (1956), *Medical Physiology*, 10th ed., C.V. Mosby, St. Louis.

Batchelor, G.K. (1970), *An Introduction to Fluid Dynamics*, Cambridge Univ. Press, New York, 148.

Bell, G., Davidson, J.N. and Scarborough, H. (1965), *Textbook of Physiology and Biochemistry*, 6th ed., E & S Livingston, Edinburgh.

Bellhouse, B.J. and Talbot, L. (1969), The fluid mechanics of the aortic valve, *J Fluid Mech*, 35, 721.

Benditt, E.P. and Benditt, J.M. (1973), Evidence for a monoclonal origin of human atherosclerotic plaques, *Proc Nat Acad Sci USA*, 70, 1753.

Benedict, J.V., Harris, E.H. and Von Roseberg, E.U. (1970), An analytical investigation of the cavitation hypothesis of brain damage, ASME Paper 70-BHF-3.

Berne, R.M. and Levy, M.N. (1972), *Cardiovascular Physiology*, C.V. Mosby, St. Louis.

Blumenthal, H.T. (1967), Hemodynamic factors in the etiology of artheriosclerosis, in *Cowdry's Artheriosclerosis*, 2nd ed., Blumenthal, H.T., ed., C.C. Thomas, Springfield, Illinois, 510.

Boussinesq, M.J. (1868), Memoire sur l'influence des frottements dans les mouvements reçuliers des fluids, *J Math Pures Appl*, ser. 2, 13, 377.

Boussinesq, M.J. (1872), Influence de forces centrifuges sur le mouvement perm. varie de l'eau dans les canauz larges, *Soc Philom Bull*, 8, 77.

Brech, R. and Bellhouse, B.J. (1973), Flow in branching vessels, *Cardiovas Res*, 7, 593.

Carlsten, A. and Grimby, G. (1966), *The Circulatory Response to Muscular Exercise in Man*, C.C. Thomas, Springfield, Illinois.

Caro, C.G. (1966), The dispersion of indicator flowing through simplified models of the circulation and its relevance to velocity profile in blood vessels, *J Physical*, 185, 501.

Caro, C.G. (1973), Transport of material between blood and wall in arteries, in *Atherogenesis: Initiating Factors*, CIBA Foundation Symposium 12 (new series), Porter, R. and Knight, J., eds., Associated Scientific Press, Amsterdam, 127.

Caro, C.G., Fitz-Gerald, J.M. and Schroter, R.C. (1971), Atheroma and arterial wall shear. Observation, correlation and proposal of a shear dependant mass transfer mechanism for atherogenesis, *Proc Roy Soc Lon*, Ser. B, 177, 109.

Cassanova, R.A., Giddens, D.B. and Mabon, R.F. (1975), A comparison of stenotic fluid dynamics in steady and pulsatile flow, 1975 ASME Biomechanics Symposium, New York, 27.

Chandran, K.B., Swanson, W.M. and Ghista, D.N. (1974), Oscillatory flow in thin-walled curved elastic tubes, *Ann Biomed Eng*, 2, 392.

Constantinides, P. (1965), *Experimental Atherosclerosis*, Elsevier, New York, 14.

Cox, R.H. (1969), Comparison of linearized wave propagation models for arterial blood flow analysis, *J Biomech*, 2, 251.

Cox, R.H. (1970), Wave propagation through a Newtonian fluid contained within a thick-walled, viscoelastic tube: The influence of wall compressibility, *J Biomech*, 3, 317.

Crow, W.J. (1969), Studies of arterial branching models using flow birefringence, Ph.D. Thesis, Univeristy of Florida, Florida.

Cuming, H.G. (1952), The secondary flow in curved pipes, *Gr Brit Aero Res Counc Reports and Memoranda*, No. 2880.

Daly, B.J. (1976), A numerical study of pulsatile flow through stenosed canine femoral arteries, *J Biomech*, 9, 465.

Davis, W. and Fox, R.W. (1967), An evaluation of the hydrogen bubble technique for the quantitative determination of fluid velocities within clear tubes, *J of Basic Eng, Trans ASME*, 89, 771.

Dean W.R. (1927), Note on the motion of fluid in a curved pipe, *Phil Mag*, 4, 208.

Dean, W.R. (1928), The streamline motion of fluid in a curve pipe, *Phil Mag*, 5, 674.

Deshpande, M.D., Giddens, D.P. and Mabon, R.F. (1976), Steady laminar flow through modelled vascular stenoses, *J Biomech*, 9, 165.

Deutsch, S. and Phillips, W.M. (1971), The use of the Taylor-Couette stability problem to validate a constitutive equation for blood, *Biorheology*, 14, 253.

Downie, H.G., Murphy, E.A., Rowsell, H.C. and Mustard, J.F. (1963), Extracorporeal circulation: A device for the quantitative study of thrombus formation in flowing blood, *Circ Res*, 12, 441.

Duncan, L.E. (1963), Mechanical factors in the localization of atheromata, in *Evaluation of the atherosclerotic plaque*, Jones, R.J., ed., University of Chicago Press, Chicago, 171.

Ehrlich, L.W. and Friedman, M.H. (1977), Steady convective diffusion in a bifurcation, *IEEE Trans Biomed Eng*, BME-24, 12.

Ellard, D., Huth, C. and Scott, A. (1968), A biomedical heat exchanger for localized cooling of the brain, Unpublished-Design project for Mec.E. 463, U of Alberta, Edmonton.

Eustice, J. (1910), Flow of water in curved pipes, *Proc Roy Soc Lon*, ser. A, 84, 107.

Eustice, J. (1911), Experiments on streamline motion in curved pipes, *Proc Roy Soc Lon*, ser. A, 85, 119.

Evans, R.L., Hosie, K.F., Kooiker, R.H., Perry, J. and Stish, R.J. (1960), Reflections in model and arterial pulse waves, *J Appl Physiol*, 15, 258.

Ferguson, G.G. and Roach, M. (1972), Flow conditions at bifurcations as determined in glass models, with reference to the focal distribution of vascular lesions, in *Cardiovascular Fluid Dynamics*, vol. 2, Bergel, D.H., ed., Academic Press, London, 141.

Fernandez, R.C., DeWitt, K.J. and Botwin, M.R. (1976), Pulsatile flow through a bifurcation with applications to arterial disease, *J Biomech*, 9, 575.

Forstram, R.J., Blackshear, P.L. Jr., Dorman, F.D., Kreid, D.K. and Kihara, K. (1971), Experimental study of a model blood flow in channels, ASME Paper 71-WA/BHF-5.

Fox, J.A. and Hugh, A.E. (1966), Localization of atheroma: A theory based on boundary layer separation, *Brit Heart J*, 28, 388.

Friedlander, S.K. and Johnstone, H.F. (1957), Deposition of suspended particles from turbulent gas streams, *Ind Eng Chem*, 49, 1151.

Fry, D.L. (1968), Acute vascular endothelial changes associated with increased blood velocity gradients, *Circ Res*, 22, 165.

Fry, D.L. (1969), Certain histological and chemical responses of the vascular interface to acutely induced mechanical stress in the aorta of the dog, *Circ Res*, 24, 93.

Fry, D.L. (1973), Response of the arterial wall to certain physical factors, in *Atherogenesis: Initiating Factors*, CIBA Foundation Symposium 12 (new series), Porter, R. and Knight, J., eds., Associated Scientific Press, Amsterdam, 93.

Fry, D.L., Mallos, A.J. and Casper, A.G.T. (1956), A catheter tip method for measurement of the instantaneous aortic blood velocity, *Circ Res*, 4, 627.

Fung, Y.C. (1969), Blood flow in the capillary bed, Prepared for a General Lecture at the Technical Conference on Biomechanics, U. of Michigan, Ann Arbor, Michigan.

Fung, Y.C. (1971), Biomechanics: A survey of the blood flow problem, in *Advances in Applied Mechanics*, vol. 11, Yih, C.S., ed., Academic Press, New York, 65.

Geer, J.C. and McGill, H.C. (1967), The evolution of the fatty streak, in *Atherosclerotic Vascular Disease*, Brest, A.N. and Moyer, J.H., eds., Appleton-Century-Crafts, New York, 8.

Gerrard, J.H. (1971), An experimental investigation of pulsating turbulent water flow in a tube, *J Fluid Mech*, 46, 43.

Gilman, S.F. (1955), Pressure losses in divided-flow fittings, *Heating Piping and Air Cond*, 27, part 1, no. 4, 141.

Glagov, S. (1972), Hemodynamic risk factors: Mechanical stress, mural architecture, medical nutrition and the vulnerability of arteries to atherosclerosis, in *The Pathogenesis of Atherosclerosis*, Wissler, R.W. and Geer, J.C., eds., William & Wikins, Baltimore, 164.

Goldsmith, H.L. (1972), The flow of model particles and blood cells and its relation to thrombogenesis, in *Progress in Hemostasis and Thrombosis*, vol. 1, Spaet, T.H., ed., Grune and Stratton, New York, 97.

Gorman, J. (1977), A running argument, *The Sciences*, 17, no. 1, 10.

Gosman, A.D., Vlachos, N.S. and Whitelaw, J.H. (1975), Measurement and calculation of laminar flow downstream of a bend in a round tube and the prognosis for similar investigations of blood flow in venules, ASME Paper 75-APMB-7.

Greenfield, H. (1969), The design of artificial heart valves and other medical problems by computer, *Data Sheet*, Official publication of the College of Engineering, University of Utah.

Greenfield, J.C. Jr. and Fry, D.L. (1962), Measurement errors in estimating aortic blood velocity by pressure gradient, *J Appl Physiol*, 17, 1013.

Guest, M.H., Bond, T.P., Cooper, R.G. and Derrick, J.R. (1963), Red blood cells: Change in shape in capillaries, *Science*, 142, 1319.

Gutstein, W.H. and Schneck, D.J. (1967), In vitro boundary layer studies of blood flow in branched tubes, *J Atheroscler Res*, 7, 295.

Gutstein, W.H., Schneck, D.J. and Marks, J.O. (1968), In vitro studies of local blood flow disturbance in a region of separation, *J Atheroscler Res*, 8, 381.

Guyton, A.C. (1960), *Function of the Human Body*, W.B. Saunders, Philadelphia.

Guyton, A.C. (1971), *Textbook of Medical Physiology*, 4th ed., W.B. Saunders, Philadelphia.

Hagen, G. (1839), Über die bewegung des wassers in engen zylindrischen, *Röhren Pogg Ann*, 46, 423.

Hardung, V. (1962), Propagation of pulse waves in visco-elastic tubings, in *Handbook of Physiology: Circulation*, Sec. 2, vol. 1, Hamilton, W.F. ed., American Physiological Society, Washington D.C., 107.

Hawthorne, W.R. (1951), Secondary circulation in fluid flow, *Proc Roy Soc Lon*, ser. A, 206, 374.

Hoeber, T.W. and Hochmuth, R.M. (1970), Measurement of red cell modulus of elasticity by in vitro and model cell experiments, ASME Paper 70-BHF-4.

Huang, H.K. (1970), Theoretical analysis of flow patterns in single-file capillaries, ASME Paper 70-BHF-10.

Hugh, A.E. and O'Malley, A.W. (1975), Correlation of intra-arterial contrast stasis with flow patterns at constrictions, branches and bends: An experimental model, *Clin Radiol*, 26, 505.

Ito, H. (1950), Theory on laminar flows through curved pipes of elliptic and rectangular cross-sections, *The Reports of the Inst High Speed*

Mech, Tokohu University, 1, 1.

Kandarpa, K. and Davids, N. (1976), Analysis of fluid dynamic effects on atherogenesis of branching sites, *J Biomech,* 9, 735.

Kaufman, B., Davey, T.B., Smeloff, E.A., Huntley, A.C. and Miller, G.E. (1968), Development of mechanical heart assists, ASME Paper 68-WA/BHF-4.

Keele, K.D. (1978), The life and work of William Harvey, *Endeavour,* New Series, 2, No. 3, 104.

Keulegan, G.H. and Beij, K.H. (1937), Pressure losses for fluid flow in curved pipes, *J Res Natl Bur Stand,* 18, 89.

Kiser, K.M., Falsetti, H.L., Yui, K.H., Resistarits, M.R., Francis, G.P. and Carroll, R.J. (1976), Measurements of velocity wave forms in the dog aorta, *J Fluids Eng, Trans ASME,* 98, 297.

Kline, K.A. and Allen, S.J. (1971), Deformation of red cells in shear fields, ASME Paper 71-WA/BHF-11.

Krovetz, L.J. (1965), The effect of vessel branching on haemodynamic stability, *Phys Med Biol,* 10, No. 3, 417.

Krueger, J.W. , Young, D.F. and Cholvin, N.R. (1970), An in vitro study of flow response by cells, ASME Paper 70-BHF-9.

Kuchar, N.R. and Ostrach, S. (1965), Flows in the entrance region of circular elastic tubes, FTAS/TR-65-3, Case Western Reserve University, Cleveland, Ohio.

Kuchar, N.R. and Scala, S.M. (1968), Design of devices for optimum blood flow, ASME Paper 68-DE-52.

Langhaar, H.L. (1942), Steady flow in the transition length of a straight tube, *J Appl Mech,* 9, A55.

Lee, J.S. and Fund, Y.C. (1968), Experiments on blood flow in lung alveoli models, ASME Paper 68-WA/BHF-2.

Lee, J.S. and Fung, Y.C. (1970), Flow in locally constricted tubes at low Reynolds numbers, *J Appl Mech,* 37, 9.

Lew, H.S. (1973), The dividing streamline of bifurcating flows in a two-dimensional channel at low Reynolds number, *J Biomech,* 6, 423.

Lighthill, M.J. (1972), Physiological fluid dynamics: A survey, *J Fluid Mech,* 52, 475.

Lighthill, Sir J. (1975), *Mathematical Biofluiddynamics*, Soc Indus. Appl. Math., Philadelphia.

Lyne, W.H. (1970), Unsteady flow in a curved pipe, *J Fluid Mech*, 45, 13.

Lynn, N.S., Fox, V.G. and Ross, L.W. (1970), Paper presented at 63rd Annual Meeting, AICHE, Chicago.

Malindzak, G.S. (1956), Reflection of pressure pulses in the aorta, *Med Res Eng*, 6, 4th quarter, 25.

Malindzak, G.S. and Stacy, R.W. (1965), Dynamic behaviour of a mathematical analog of the normal human arterial system, *Amer J Med Elec*, 4, no. 1, 28.

Malindzak, G.S. and Stacy, R.W. (1966), Dynamics of pressure pulse transmission in the aorta, *Ann NY Acad Sc*, 128, Art, 3, 921.

Mark, F.F., Bargeron, C.B. and Friedman, M.H. (1975), An experimental investigation of laminar flow in a rectangular cross-section bifurcation, Applied Physics Lab, John Hopkins U, Silver Spring, Maryland.

Martin, J.D. and Clark, M.E. (1966), Theoretical and experimental analyses of wave reflections in branched flexible conduits, *Eng Mech Res*, ASCE, 441.

McDonald, D.A. (1955), The relation of pulsatile pressure to flow in arteries, *J Physiol*, 127, 533.

McGill, H.C., Geer, J.C. and Strong, J.P. (1963), Natural history of human atherosclerotic lesions, in *Atherosclerosis & Its Origin*, Sandler, M. and Bourne, G.H., eds., Academic Press, New York, 39.

Medical College, New York Hospital (1970) Filtering blood may avert bypass brain damage, *JAMA*, 212, 1450.

Meisner, J.E. and Rushmer, R.F. (1973), Eddy formation and turbulence in flowing liquids, *Circ Res*, 12, 455.

Meisner, J.E. and Rushmer, R.F. (1968), Production of sounds in distensible tubes, *Circ Res*, 12, 651.

Mitchell, J.R.A. and Schwartz, C.J. (1965), The localization of arterial plaques, in *Arterial Disease*, Blackwell Scientific Publications, Oxford 50.

Mueller, T.J., Llyod, J.R. Chetta, G.E. and Galanga, F.L. (1975), Effect of test section geometry on the occluder motion of caged-ball prosthetic heat valves, 1975 ASME Biomechanics Symposium, New York, 55.

Mullinger, R.N. and Manley, G. (1967), Glycosaminoglycans and atheroma in iliac arteries, *J Atheroscler Res*, 7, 401.

Nerem, R.M. and Seed, W.A. (1972), An in vivo study of aortic flow disturbances, *Cardiovasc Res*, 6, 1.

O'Brien, V., Ehrlich, L.W. and Friedman, M.H. (1976) Unsteady flow in a branch, *J Fluid Mech*, 75, 315.

Park, S.D. and Lee Y. (1971), Diabatic turbulent flow in the entrance region of concentric annuli, *Eng J*, 54, No. 6.

Patel, D.J. and Janicki, J.S. (1966), Catalogue of some dynamic analogies used in pulmonary and vascular mechanics, *J Med Res Eng*, 5, 30.

Pedley, T.J., Schroter, R.C. and Sudlow, M.F. (1971), Flow and pressure drop in systems of repeatedly branching tubes, *J Fluid Mech*, 46, 365.

Pelot, R.P. and Rodkiewicz, C.M. (1982), Aortic Arch Separation and Flow Patterns - In Vitro Study, Dept. Report No. 27, Dept. of Mech. Eng., University of Alberta, Edmonton.

Poiseuille, J. (1840, 1841), Recherches expérimentalles sur le mouvement des liquids dans les tubes de tres-petits diamétres, *Comptes Rendus*, 11, 961, 1041, and 12, 112.

Porgé, I.G. (1964), Hemodynamics of the ascending aorta, in *Pulsatile Blood Flow*, Attinger, E.O., ed., McGraw Hill, New York, 237.

Reif, T.H., Nerem, R.H. and Kulacki, F.A. (1976), An in vitro study of transendothelial albumin transport in a steady state pipe flow at high shear rates, *J Fluid Eng, Trans* ASME, 98, 488.

Resch, J.A., Okabe, N., Loewenson, R.B., Kimoto, K., Katsuki, S. and Baker, A.B. (1969), Pattern of vessel involvement in cerebral atherosclerosis, *J Atheroscler Res*, 9, 239.

Rodbard, S. (1956), Vascular modifications induced by flow, *Amer Heart J*, 51, 926.

Rodkiewicz, C.M. (1974), Atherosclerotic formations in the light of fluid mechanics, *Proc 1974 Fluids Eng Conf*, Montreal.

Rodkiewicz, C.M. (1974), Bifurcation characteristic Reynolds number, *J Hydraulic Res*, 12, 241.

Rodkiewicz, C.M. (1975), Localization of early atherosclerotic lesions in the aortic arch in the light of fluid flow, *J Biomech*, 8, 149.

Rodkiewicz, C.M. and Hung, R. (1976), Flow division dependence of some large arterial junctions on frequency and amplitude, Departmental

Report No. 5, Dept. of Mech. Eng., U of Alberta, Edmonton, Alberta.

Rodkiewicz, C.M. and Kalita, W. (1978), Flow characteristics of the curved pipe junctions, ·Departmental Report No. 12, Dept. of Mech. Eng., U of Alberta, Edmonton, Alberta.

Rodkiewicz, C.M. and Roussel, C.L. (1973), Fluid mechanics in a large arterial bifurcation, *J Fluids Eng, Trans* ASME, 95, 108.

Rodkiewicz, C.M. and Wong, S.L. (1974), On the mass flow division in curved junctions, *ASHRAE Trans*, 80, Part 2, 280.

Rogers, V.A. and Moskowitz, G.D. (1971), Analysis for hydrodynamic model of human systemic arterial circulation, ASME Paper 71-WA/AUT-13.

Rosenhear, L. (1973), *Laminar Boundary Layers*, Oxford University Press, London.

Roussel, C.L. and Rodkiewicz, C.M. (1973), Edge effect in large arterial bifurcation, *J Fluids Eng, Trans* ASME, 95, 334.

Rowe, M. (1970), Measurements and computations of flow in pipe bends, *J Fluid Mech*, 43, 771.

Sato, Y. (1963), Separating and uniting flows in a branch pipe, *J Japan Soc Mech Eng*, 66, No. 537.

Scarton, H.A., Shah, P.M. and Tsapogas, M.J. (1975), The role of hemodynamics in early atheroma in the aorta, 1975 ASME Biomechanics Symposium, New York, 15.

Scarton, H.A., Shah, P.M. and Tsapogas, M.J. (1977), Relationship of the spatial evolution of secondary flow in curved tubes to the aortic arch, in *Mechanics in Engineering*, Dubey, R.N. and Lind, N.C., eds., U of Waterloo Press, Waterloo, Canada, 111.

Scherer, P.W. (1973), Flow in axisymmetrical glass model aneurysms, *J Biomech*, 6, 695.

Schlichting, H. (1968), *Boundary Layer Theory*, 6th ed., McGraw Hill, New York.

Schneck, D.J. (1977), Pulsatile blood flow in a channel of small exponential divergence, III - Unsteady flow, *J Fluids Eng, Trans* ASME, 99, 33.

Schneck, D.J. and Gustein, W.H. (1966), Boundary-layer studies in blood flow, ASME Paper 66-WA/BHF-4.

Schraub, F.A., Kline, S.J., Henry, J., Runstadler, P.W. Jr. and Littell,

A. (1965), Use of hydrogen bubbles for quantitative determination of time dependant velocity fields in low-speed water flows, J Basic Eng, Trans ASME, 87, 429.

Schultz, D.L. (1972), Pressure and flow in large arteries, in Cardiovascular Fluid Dynamics, vol. 1, Bergel, D.H. ed., Academic Press, London, 281.

Seed, W.A. and Wood, N.B. (1971), Velocity patterns in the aorta, Cardiovas Res, 5, 319.

Shah, P.M., Tsapogas, M.J. Scarton, H.A., Jindal, P.K. and Wu, K.T. (1976) Predilection of occlusive disease for the left iliac artery, J Cardiovasc Surg, 17, 420.

Sharma, M.G. and Hollis, T.M. (1976), Rheological properties of arteries under normal and experimental hypertensive conditions, J Biomech, 9, 293.

Singh, M.P. (1974), Entry flow in a curved pipe, J Fluid Mech, 65, 517.

Skalak, R. and Branemark, P.I. (1969), Deformation of red blood cells in capillaries, Science, 164, 717.

Snyder, M.F., Rideout, V.C. and Hillestad, R.J. (1968), Computer modelling of the human systemic arterial tree, J Biomech, 1, 341.

So, R.M.C. (1976), Entry flow in curved channels, J Fluid Eng, Trans ASME, 98, 305.

Spain, D.M. (1966), Atherosclerosis, Sc. Am, 215, 49.

Spencer, M.P. and Denison, A.B. (1956), The aortic flow pulse as related to differential pressure, Circ Res, 4, 476.

Stehbens, W.E. (1974), Changes in the cross-sectional area of the arterial fork, Angiology, 25, 561.

Stehbens, W.E. (1974), Hemodynamic production of lipid deposition, intimal tears, mural dissection and thrombosis in the blood vessel wall, Proc Roy Soc Lon, ser. B, 185, 357.

Strehler, E. and Schmid, P. (1970), Nomogram for determining normal aortic diameter (aortic arch) and aortic biological age in 2-m chest x-rays, in Scientific Tables, 7th ed., Diem, K. and Lentner, C., eds., J.R. Geigy S.A., Basel, Switzerland.

Szmidt, E.W., Bieliczenko, J.A., Bogatyriew, J.W., Kniaziew, M.D. and Pokroweskij, A.W. (1973), The Obliterative occlusions of the carotid arteries and their surgical treatment (in Russian), Khrirurgiia, 49, 3.

Taylor, G.I. (1929), The criterion for turbulence in curved pipes, *Proc Roy Soc Lon*, Ser. A, 124, 243.

Taylor, M.G. (1957), An approach to an analysis of the arterial pulse wave, *Phys Med Biol*, 1, 258.

Taylor, M.G. (1963), Wave travel in arteries and the design of the cardiovascular system, in *Pulsatile Blood Flow*, Attinger, E.O., ed., McGraw Hill, New York, 343.

Texon, M. (1963), The role of vascular dynamics in the development of atherosclerosis, in *Atherosclerosis and Its Origin*, Sandler, M. and Bourne, G.H., eds., Academic Press, New York, 167.

Texon, M. (1967), Mechanical factors involved in atherosclerosis, in *Atherosclerotic Vascular Disease*, Brest, A.N. and Moyer, J.H., eds., Appletcn-Century-Crafts, New York, 23.

Texon, M. (1972), The hemodynamic basis of atherosclerosis, further observations: The ostial lesson, *Bull N Y Accd Med*, 48, 733.

Texon, M. (1974), Atherosclerosis: Its hemodynamic basis and implications, *Symposium on Atherosclerosis*, Medical Clinics of North AMerica, 58, No. 2, 257.

Thompson, J. (1876), On the origin of windings of rivers in alluvial plains with remarks on the flow of water round bends in pipes, *Proc Roy Soc Lon*, 25, 5.

Thompson, J. (1877), Experimental demonstration in respect to the origin of winding of rivers in alluvial plains, and to the mode of flow of water round bends of pipes, *Proc Roy Soc Lon*, 26, 356.

Thompson, J. (1878), On the flow of water in uniform régime in rivers and other open channels, *Proc Roy Soc Lon*, 28, 114.

Thompson, J. (1879), Flow round river bends, *Prov Inst Mech Engrs*, 2, 456.

Thurston, G.B. (1975), Viscoelastic resonance and impedance for blood flow in large tubes, 1975 ASME Biomechanics Symposium, New York, 39.

Timm, C. (1942), Der strömungsverlauf in einem modell der menschichen aorta, *Z Biol*, 101, 79.

Turnstall, M.J. and Harvey, J.K. (1968), On the effect of a sharp bend in a fully developed turbulent pipe flow, *J Fluid Mech*, 34, 595.

Tuttle, W.W. and Schottelius, B.A. (1965), *Textbook of Physiology*, 15th ed., C.V. Mosby, St. Louis.

Uchida, S. (1956), The pulsating viscous flow superposed on the steady laminar motion of incompressible fluid in a circular pipe, Z Angewandte Math Physics, 7, 403.

Vazsonyi, A. (1944), Pressure loss in elbows and duct branches, Trans ASME, 66, 177.

Walton, K.W. and Williamson, N. (1968), Histological and immunofluorescent studies on the evolution of the human atheromatous plaque, J Atheroscler Res, 8, 599.

Wells, H.K., Winter, D.C., Nelson, A.W. and McCarthy, T.C. (1977), Blood velocity patterns in coronary arteries, J Biomech Eng, Trans ASME, 99, 26.

Wesolowski, S.A., Fries, C.C., Sabini, A.M. and Sawyer, P.N. (1962), The significance of turbulence in hemic systems and in the distribution of the atherosclerotic lesion, Surgery, 57, 155.

Wesolowski, S.A., Fries, C.C., Sabini, A.M. and Sawyer, P.N. (1965), Turbulence, intimal injury and atherosclerosis, in Biophysical Mechanisms in Vascular Homeostasis and Intravascular Thrombosis, Sawyer, P.N., ed., Appleton-Century-Crofts, New York, 147.

White, C.M. (1929), Streamline flow through curved pipes, Proc Roy Soc Lon, ser, A, 123, 645.

White, F.M. (1974), Viscous Fluid Flow, McGraw Hill, New York.

Whitmore, R.L. (1968), Rheology of the Circulation, Pergamon Press, New York.

Wieting, D.W. (1968), A method of analysing the dynamic flow characteristics of prosthetic heart valves, ASME Paper 68-WA/BHF-3.

Wieting, D.W., Akers, W.W., Feola, M., and Kennedy, J.H. (1970), Analysis of a variable volume intra-aortic balloon pump in a mock circulatory system, ASME Paper 70-BHF-5.

Wintrobe, M.M. (1967), Clinical Hematology, Lea and Febiger, Philadelphia.

Wormersly, J.R. (1955), Method for the calculation of velocity, rate of flow and viscous drag in arteries when the pressure gradient is known, J Physiol, 127, 553.

Yao, L.W. and Berger, S.A. (1975), Entry flow in a curved pipe, J Fluid Mech, 67, 177.

Yellin, E.L., Peskin, C.S. and Frater, R.W.M. (1972), Pulsatile flow across the mitral valve: Hydraulic, electronic and digital computer simulation, ASME Paper 72-WA/BHF-10.

Young, D.F. and Tsai, F.Y. (1972), Flow characteristics in models of arterial stenoses, I. - Steady flow, II. - Unsteady flow, J *Biomech*, 6, 395 and 547.

Zareckij, W.W., Sandrikow, W.A., Wychowskaja, A.G., Sablin, J.N. and Lemieniew, J.N. (1973), The study of the blood flow in the pathologic structures of the peripheral blood vessels (in Russian), *Khrirurgiia*, 49, 24.

Zechmeister, A. (1969), Calcification of epicardial stretches of bridged coronary arteries in man, J *Atheroscler Res*, 9, 121.

CONTRIBUTORS

The experience and background of the contributors differs widely
and a brief biography of each is provided to give one an awareness of
the qualifications of the authors.

Daniel Quemada of the Universite Paris 7, France, contributed
Chapter I. He is full professor of Physical Hydrodynamics and Biomech-
anics at the University of Paris VII and head of the "Laboratory of
Biorheology and Physico-Chemical Hydrodynamics" sponsored by the C.N.R.S.

He is a former student of the Ecole Normale Superiéure, one of the
most prestigious centers of high education in France, where he completed
its curriculum in 1954. He owns, since 1964, a Ph. D. (Docteur d'Etat)
on plasma physics from the University of Orsay and became Assistant
Professor in the Physics Department of the University of Paris in 1966.
Between 1964 and 1972, he conducted research on plasma dynamics in the

outer space with special interest in MHD models of magnetosphere. As a result of this work, he authored a book on "Waves in Plasma" (in French) and a number of scientific papers on the subject.

Since 1971, he directed his interest towards problems related with biorheology and biomechanics and developed a research group whose contributions to this matter has become recognized worldwide.

He is a member of the board of editors of Biorheology, of the British Society of Rheology, of the Société Francaise de Physique and of the Société de Chimie-Physique. His contribution to the field of blood flow comprises more than twenty papers treating from the pulsatory flow in large vessels to the modelization of microflow of dispersed systems of deformable particles.

Joseph Barbenel of the University of Strathclyde, Scotland, contributed Chapter II. He received his B.D.S. London, L.D.S., R.C.S. (Eng.) in 1960 with a prize in Anatomy and Dental Anatomy. From 1960 to 1962 he served as Lt. R.A.D.C. and later Capt. R.A.D.C. in the National Service in Malaya. During those two years, he provided a complete service for Army personnel and families. In addition, he was in charge of the dental departments of BMH, Kinrara and BMH, Kamunting for approximately 7 months.

Dr. Barbenel joined the University of St. Andrews in Dundee as a student in 1963, and graduated in 1966 with a B.Sc., receiving a Junior Honours Physics Medal as well. He became a student at the Bioengineering Unit of the University of Strathclyde in 1966, and in 1967 received his M.Sc., (Strathclyde) with a MRC award for Further Education in the

Medical Sciences. From 1967 to 1969 he was a lecturer at the Dept. of Dental Prosthetics, Dental School and Hospital, University of Dundee.

Dr. Barbenel became Senior Lecturer at the Bioengineering Unit, of Stratclyde in 1970, and in 1973 he was Head of Tissue Mechanics Division. In that year he also became a M. Inst.P. He received his Ph.D. (Strathclyde) in 1979. From November to December 1979, he was visiting Professor at the National Engineering University, Lima, Peru.

Jim B. Haddow of the University of Alberta, Canada, is the principal author of Chapter III. He took his B.Sc. at the University of St. Andrews, Scotland, in 1951 and his Ph.D. degree in Engineering at the University of Manchester in 1960. From 1955 - 58 and from 1961 to the present he has been a member of the Mechanical Engineering Department at the University of Alberta except for one year as a visiting professor at the University of New South Wales.

The co-authors with Dr. Haddow are B. Moodie and R.J. Tait, both also with the University of Alberta.

Bryant Moodie received his B.Sc. degree in mathematics from Carlton University, Ottawa and his M.Sc. and Ph.D. degrees in mathematics from the University of Toronto. From 1970 - 72 he was a research associate at Carleton University, from 1972 - 73 a visiting lecturer at the University of Dundee, Scotland and is presently Associate Professor of Mathematics at the University of Alberta, Edmonton and recipient of a McCalla Research Professorship.

R.J. Tait took his Ph.D. degree in applied mathematics at Glasgow with I.N. Sneddon in 1962. He joined the department at Edmonton the same year and apart from a year at Norman, Oklahoma, and another at the

Mathematics Institute at Warwick is, at present, still there.

Thomas Kenner of the Physiologisches Institut der Universität Graz

contributed chapter IV. He is full professor and chairman of the phys-

iological institute in Graz. He received his M.D. in 1956 at the Uni-

versity of Vienna, Austria, and has since that time performed clinical

studies and experimental and theoretical research on various aspects of

applied and theoretical hemodynamics. Together with E. Wetterer he pub-

lished a monography on the arterial pulse in 1968. From 1968 to 1971

he performed research on circulatory control at the division of biomed-

ical engineering of the University of Virginia. In 1972, he was appo-

inted full professor and chairman at the position he still holds in

Graz. His interest is currently directed towards the relation between

dynamics and control of blood flow, towards the analysis of stochastic

phenomena in the circulation, and towards the application of a new

method for the continuous measurement of blood density for the analysis

of fluid exchange in the microcirculation.

Czelaw M. Rodkiewicz of the University of Alberta, Canada, contrib-

uted Chapter V. He is a full professor in the Mechanical Engineering

Department and has conducted research on various aspects of blood flow.

He received his Diploma in Mechanical Engineering at the Polish Univer-

sity College in London, Great Britain, and his M.Sc. degree at the

University of Illinois, U.S.A. He earned his Ph.D. degree at the Case

Institute of Technology in Cleveland, Ohio, U.S.A.

Dr. Rodkiewicz has done extensive research into fluid mechanics and heat transfer, specifically in the domains of hypersonic flight, lubrication and blood flow. He has written numerous papers on the subject and given lectures at various conferences and invited lectures and tours in many countries. Dr. Rodkiewicz is listed in the "Who's Who in the World" and many other biographical books.

Printed in the United States
By Bookmasters